Making Political Geography

Making Political Geography

Second Edition

John Agnew and Luca Muscarà

ROWMAN & LITTLEFIELD PUBLISHERS, INC.
Lanham • Boulder • New York • Toronto • Plymouth, UK

Published by Rowman & Littlefield Publishers, Inc.
A wholly owned subsidiary of The Rowman & Littlefield Publishing Group, Inc.
4501 Forbes Boulevard, Suite 200, Lanham, Maryland 20706
www.rowmanlittlefield.com

Estover Road, Plymouth PL6 7PY, United Kingdom

British Library Cataloguing in Publication Information Available

Library of Congress Cataloging-in-Publication Data

Agnew, John A.
Making political geography / John Agnew and Luca Muscarà. — 2nd ed.
p. cm.
Includes bibliographical references and index.
ISBN 978-1-4422-1229-9 (cloth : alk. paper) — ISBN 978-1-4422-1230-5 (pbk. : alk. paper) — ISBN 978-1-4422-1231-2 (electronic)
1. Political geography. 2. Geopolitics. I. Muscarà, Luca. II. Title.
JC319.A44 2012
320.1'2—dc23

2011045007

™ The paper used in this publication meets the minimum requirements of American National Standard for Information Sciences—Permanence of Paper for Printed Library Materials, ANSI/NISO Z39.48-1992.

Printed in the United States of America

Contents

Preface

The Palazzo Pubblico or town hall in Siena, Italy, contains one of the best-known fourteenth-century Italian fresco cycles. By native son Ambrogio Lorenzetti (c. 1295–1348), these are predominantly secular paintings in that they portray sequentially around the walls of a single room the artist's imaginative rendering of the consequences in city and country of good and bad government. In the words of Erwin Panofsky (1969, 142), a leading twentieth-century art historian, Lorenzetti's city and countryside were "the first post-classical vistas essentially derived from visual experience rather than from tradition, memory, and imagination." But there is also some serious idealization at work in these images: "'reflecting' and 'seeing' the ideal city republic" (Starn 1994, 31). They are thus visual statements about both how the world usually works politically and how it should do so. The series of frescoes could be regarded as one of the earliest attempts at "making political geography," both analytically in terms of what has been achieved and normatively about what can go wrong if the mode of government is corrupted.

Political geography is about understanding the geographical distribution of power, how it concentrates in some hands and some places, the human and environmental consequences of such concentration, and how it shifts between places over time. Wars have been one of the ways in which power is redistributed, not often to the benefit of those who start them. All manner of conflicts short of open warfare and the various mechanisms that people have devised down the years to manage conflict and provide public goods, what is usually thought of as "politics," can also lead to significant shifts in the use and manifestation of power. Lorenzetti's frescoes do not just carry some sort of transcendental lesson for the ages about every one of these aspects of "the political." Their very existence points to ways in which the political intrudes into how we represent places such as Siena and time periods such as the European fourteenth century. If we focus solely on the magnificence of the art and its "civic showmanship," we will miss the fact that Lorenzetti's paintings were an effort to recoup Siena's political reputation after military defeat and as "its merchants and bankers were losing to the competition, and its territorial boundaries, if not its ambitions, had receded" (Starn 1994, 13). Art can be a political weapon, albeit one that while bolstering your identity can, when the art is brilliant enough, be expropriated by others. Down the years one wonders how many non-Sienese upon encountering Lorenzetti's work in the Sala dei Nove in the Palazzo Pubblico have read their own (national) stories into the frescoes.

This understanding of the historical and geographical contingency of political identities and the stories that are woven together to give them meaning is not one that has always prevailed in the study of "political geography." Political geographers from the start had trouble with the "primacy of the political" as it is implied in Lorenzetti's paintings. They have usually been happier mining for some sort of environmental or economic determinism within the term *geography* and seeing politics as a by-product of something more fundamental "in the last instance," at least. As a modern academic field of study, political geography came into being across Europe in the 1890s as an "aid to statecraft" on the part of the Great Powers of the day. Contemporary national political identities and reason-of-state were taken as givens. The "needs" of territorial states and their relative locations on the earth's surface and the resources available in driving and determining the outcome of competition between them were the main concerns of the field. Lurking close beneath the surface of late nineteenth-century political geography and the thinking of its founders such as Ratzel and Mackinder was the view that nature determines the form of political institutions such as states and empires. If the idea of "levels of development" associated with different national-state territories came from eighteenth-century Scottish Enlightenment thought, that of the hierarchies of such territories emerging from competition with one another for domination came from nineteenth-century German Idealism. This intellectual genealogy predominated down until the 1940s and still has adherents. In fact, until very recently, and reflecting the naturalistic undertone of the field in its origins, most political geographers have tended to be either imperialists (advocates of particular empires that would combine the racial-natural attributes of different environmental niches within a single centralized polity) or cameralists of a sort (advocates of state-based territorial economies that would recapitulate within their borders the environmental variety of the world at large). Not only right-wing nationalists but also left-wing populists active in political geography have frequently based their understandings of how the world should work on the latter position: believing that there is a collective interest defined by coresidence within the same territory and that economic development mandates the mercantilist view that transactions with other territories should be minimized to the greatest extent possible. Since the 1970s greater pluralism has prevailed with liberal perspectives, pitting markets against states, and social perspectives, looking to a plurality of forms of governance, becoming much more influential than the older views.

But it is not only the range of implicit political projects or normative positions informing the field that have changed. The general intellectual orientation has shifted one hundred and eighty degrees. Since the 1960s a more independent and critical approach has begun to develop, acknowledging the need to question rather than actively serve the particular interests of the "home state" of the political geographer. At the same time, the empirical scope of the field has widened to consider questions about the origins and spread of political movements, the links between places and identities, and geographies of nationalism and ethnic conflict. We have gone from "states are everything" (particularly *my* state) to "the political is everywhere." This empirical expansion of the field has transformed political geography from a peculiarly state-obsessed field to one concerned with the ways in which "the political" broadly construed intersects with geographical patternings: from the material and discursive construction of states and other polities and their interrelations to the connections that can be drawn between places and political identities.

The geopolitical context of the time has been crucial to the making of academic political geography over the past one hundred years. This is the basic premise and leitmotif of this book. The field has not evolved simply as the result of an internal dynamic, as one "paradigm" simply replaced another because of intellectual fancy or academic competition. It is not that such considerations have been absent. But they have been relatively less salient to the making of the field than the nature of the world that political geography has claimed to directly report on and interpret. The time of modern political geography's founding in the 1890s was one of burgeoning interimperial rivalry between a set of Great Powers—Germany, Britain, France, the United States, Russia, and Japan—that reached its twin peaks in the two World Wars. This period gave rise to the political geography that privileged the role of physical geography in determining or conditioning state prospects and limits. The Cold War of 1945–1989, with its emphasis on global ideological competition between two models of "modernity"—the democratic capitalism of the United States and the state socialism of the Soviet Union—initially produced a diminished interest in the study of political geography. The field as it had existed before World War II did not seem to offer much in the new circumstances. Of course, the period did encourage a "freezing" of political boundaries and a seemingly permanent standoff between the two sides. As the Cold War slowly eroded, however, political geography underwent something of a revival in the United States and elsewhere as what had seemed frozen melted in unpredictable ways. The disastrous impact of the Vietnam War on the American economy and politics, the civil rights struggles in the United States, the major economic challenges facing the United States and the world economy following the collapse of the postwar economy, and the rediscovery of "the market" as a guiding political concept all conspired *beginning* in the 1970s to create what Daniel Rodgers (2011) calls an "Age of Fracture," in which old assumptions about political and economic predictability and certainty are openly called into question. Few fields have been untouched by the wrenching changes of the period beginning in the 1970s. For political geography, not only did the distinction between the "domestic" and the "foreign" begin to break down as a meaningful device for organizing thinking in the face of a globalizing world economy, but all manner of new topics emerged for study, from the regional-ethnic movements empowered by the end of the Cold War and the invocation of an oft ill-defined "neoliberalism" as the ruling ideology of the age in explaining just about anything, to the challenges facing definitions of citizenship and understandings of nationality in a world of vastly increased international migration. The chapters that follow tell the story of the making of political geography in light of dramatic shifts in the external environment facing the field both in the dimmer past and in the more immediate present.

At the time of writing the first edition of this book, the attacks by agents of the militantly Islamist al-Qaeda network on the New York Trade Center and the Pentagon building of the U.S. Department of Defense in Washington, DC, were followed by the U.S. government's declaration of war on "international" terrorism. Many Americans reacted by flying the national flag as a symbol both of their threatened identity and as a sign of their mobilization and support of a military response. These events seemed to portend the emergence of a geopolitical context in which states, however mighty, would confront shadowy networks of discontented and fanatic groups following this or that objective, often of a religious or ethnic nature. As this feature of a new world (dis)order and other dimensions of it, such as increased flows of money, goods, and people between localities and regions in it, have taken geographical shape, political geography has also changed its shape in order to deal with what has changed. More recent events—such as the financial crisis of 2007–2009 and its pointing toward a restructuring of the world economy around an ascendant Asia and a declining Europe and United States, if also with an apparently enhanced role for national governments; the emergence of global environmental challenges such as climate change, a polluted food chain, and regional water shortages; the rebirth of militant nationalism around the world; and the collapse of many erstwhile states, such as Somalia—signal a very different global political environment from that of forty years ago. Of course, not all is "change." One of the great virtues of Lorenzetti's fresco cycle lies in its story of how political life gets redefined territorially and also develops in other still very geographically predictable ways. Therefore, even in the present heavily networked world—corporate, terrorist, informational, and cultural—identities and interests continue to take reshaped but still often territorialized forms. Across the board, it looks as if political geography will not soon be short of things to study.

A number of people have helped us bring the two editions of this book to completion. Alec Murphy was the primum mobile behind the first edition. Laura McKelvie at Arnold was its editorial sponsor. With the second edition at Rowman & Littlefield, Susan McEachern has been a crucial sponsor and intermediary. For the first edition, Troy Burnett helped enormously with research on and writing up the vignettes in chapter 2, showing why political geography still matters, and with summarizing the content of textbooks from the interwar period that we still refer to in the second edition in chapter 3. Our UCLA colleague Michael Curry found a copy of the March 2000 *American Philatelist* in an airplane seat pocket and handed it over because it contained an article on the stamps of Nagorno-Karabagh, the territory disputed by Armenia and Azerbaijan in the former Soviet Union. This topic is now one of the vignettes in chapter 2. Many departmental colleagues at UCLA, particularly the late Denis Cosgrove, Nick Entrikin, Allen Scott, and Dave Rigby, provided professional support and encouragement. Chase Langford and Matt Zebrowski, the UCLA departmental cartographers at the time of the two editions, respectively, drafted many of the figures and prepared all of them for publication. For the second edition we would particularly like to thank the following: from California, Margaret "Cariadne" Mackenzie-Hooson, who offered precious sources on the history of geographical thought from the library of her late husband, the transnational geographer David Hooson; from post-Fukushima Tokyo, Yasuo and Maiko Miyakawa provided a needed understanding of the context of the 2011 Fukushima nuclear disaster; from Italy, Mariavittoria Albini assisted in research on new and old topics; from France, André-Louis Sanguin reviewed the brief portrait of André Siegfried; from Arizona, Thomas Puleo provided wonderful support across the board; while discussions with Stefano Dejak, Italy's special envoy to Somalia, enhanced our understanding of piracy and of al-Qaeda terrorism; many at the University of Molise, in Campobasso and Isernia, have in different ways sustained this project; finally, many thanks to Laura Canali and Lucio Caracciolo at *Limes*, for inspiring geopolitical charting; to David Lowenthal and Shirley Adelson, for their extraordinary support respectively on "geographers at war" and on Jean Gottmann, for the early "discovery" of whom we are grateful to Calogero Muscarà.

The changes to the book for the second edition were mapped out by the two coauthors in the shadow of Siena's Palazzo Pubblico in October 2010. The story about Lorenzetti's great frescoes, therefore, is not simply an artful ploy. Our thinking about the nature of the political in political geography does reflect our commitment to the republican ideals of politics expressed by Lorenzetti and arguably characteristic at least to a degree of medieval Siena. With two authors, the book benefits from a wider range of experience of both geopolitical context, particularly of the recent past, and what has transpired in the making of political geography since the first edition in 2002. The authors have known one another for over twenty years but bring to the book very different backgrounds and sensibilities about its subject matter. All told, therefore, the "external context" has counted as much in the writing of this book as it has in the making of political geography.

John Agnew
Luca Muscarà
11 September 2011

1

Introduction

The commonsense meaning of political geography is the study of how politics is informed by geography. For a long time this meant trying to show how the physical features of the earth—the distribution of the continents and oceans, mountain ranges, and rivers—affected the ways in which humanity divided the world up into political units such as states and empires and how these units competed with one another for global power and influence (e.g., Bassin 2003). Today the dominant meaning has changed considerably. On one side, geography is now understood as including social and economic differences between places without necessarily ascribing these to physical differences. On the other side, politics has been broadened to include questions of political identity (how social groups define themselves and their political objectives) and political movements (why this movement or political party started here and why it has this or that geographical pattern of support). Even more fundamentally, "geography" is itself often now thought of as the selection and ranking of certain themes and issues—from the naming of the continents and the division of the world into regions to the identification of certain regions as more or less "strategically" important—rather than a set of objective facts everyone knows and that are beyond dispute. In this understanding, knowledge cannot be readily separated from power. In particular, those with access to the most power, the ability to command others, are able to define what counts as geography. From this point of view, the meaning of political geography is completely reversed: it now becomes the study of how geography is informed by politics.

From this viewpoint, political geography offers an alternative way of thinking about the world than is provided by such old but increasingly tired oppositions as freedom versus order or universals versus particulars that are at the heart of contemporary political philosophy and comparative politics. It sees the political uses of space arising as politics "concretely organizes the spaces of liberty, citizenship, law enforcement, and institutional efficacy. Politics extends the spaces of domination, traces lines of exclusion, designs internal and external borders, determines the centers and peripheries, the 'highs' and the 'lows,' and articulates the spaces of production and consumption" (C. Galli 2010, 5).

We should make clear that even as this "new" understanding has become widely established, the older one still lives on. In fact physical- and economic-determinist views of political geography—politics is *always* about something else—are anything but spent forces, as we shall see later. This book is a critical and necessarily selective survey of how and why political geography has changed as it has accumulated these varied meanings. The story is broadly that of a drift from a monist or singularly dominant sense of what the field is about and how it understands its subject matter to a much more theoretically and substantively pluralist field of study. The "making" in the title, therefore, is of vital importance. It signifies the focus on the people, historical contexts, and scholarly works that have produced the various meanings of political geography down the years. Political geography is also not just a "branch" of the field of geography, as if that broader field emerged fully formed in the late 1800s with a complete set of internal divisions. It has been "made" through the ways in which the world's actual geography has been implicated in the practice of politics over the past one hundred years by all students of politics, including many who have also been influential practitioners. Political geography, then, is a set of scholarly and political ideas about the relationship of geography to politics and vice versa that have roots in a number of fields, particularly in geography and political science, but also in sociology, anthropology, ethnic studies, and international relations. The task of this book is to trace the development of these ideas, portray some of the important authors and their academic and political influence, identify some of the crucial philosophical and theoretical issues within political geography today, and suggest some of the substantive themes of emerging significance in the early twenty-first century.

Today, a whole variety of what can be considered "public issues" of great import and publicity have profoundly geographical aspects to them. Often, knowing or identifying these geographical aspects helps to make greater sense out of otherwise mysterious situations and events than does simply reciting the "facts" as they are frequently presented. This is what contemporary political geography is mainly about: deciphering the ways in which geographical considerations enter into politics of all sorts. Such considerations, however, do not usually constitute the "common sense" of much of what goes for common knowledge, history, and social science around the world. Common sense, to quote Duncan Watts (2011, 3–4), is "exquisitely adapted to handling the complexity that arises in everyday situations. . . . But 'situations' involving corporations, cultures, markets, nation-states, and global institutions exhibit a very different kind of complexity from everyday situations. And under these circumstances, common sense turns out to suffer from a number of errors that systematically mislead us." We are all too prone to accept such conventional nostrums as "all ethnic conflicts are the product of ancient hatreds" or "global politics is all about the pursuit of rationally defined 'national interests.'" The basic problem is that we project from what works for individuals (ourselves in everyday life) and the often unacknowledged conventional "wisdom" we have acquired over time to the behavior of groups and institutions, and we then further fail to see that the experiences of groups and institutions in other places can be very different from that of our own. This is why specialized fields have arisen to address these lacunae in common sense. Political geography, therefore, is a field that exists to help understanding of how the world works politically beyond the limitations of what goes for common sense.

Rather than continue in an abstract vein, it is useful to take some public issues arbitrarily from the news over the past ten years and emphasize their geographical aspects and how these figure into their very constitution as "public problems." Looking at these problems geographically gives us an extra purchase on understanding them that we otherwise would not have. The examples may also help in understanding the fascination for political geography that has drawn people to its study over the years:

- Describing the United States as divided since the 1980s into red (Republican) and blue (Democratic) states has become a basic part of the "common sense" about American national politics. The truth to it lies in the fact that in presidential elections votes are gathered by state and awarded to the party candidate with the majority in that state. Then a map can be constructed that is red and blue in terms of so-called electoral votes. But some commentators have gone much further than this in alleging that the country has become polarized geographically between the two main political parties at the state level. There is much dispute over the "reality" of geographical polarization. It is easy to show that, though some states have trended away from one party and toward the other in all sorts of elections, much of the country is "purple" when data are aggregated at the county or electoral-district level, suggesting that it is often only a few percentage points that separate candidates of the two parties, even in what can be characterized (in electoral college terms) as the deepest of red and blue states. To some writers this suggests no or only limited geographical electoral polarization. In their view, party elites and activists have become increasingly polarized, particularly in the case of the Republican Party's ideological drift to the right. The polarization of American politics is not a geographical/popular phenomenon at all but only apparently geographical because voters increasingly face more politically polarized candidates than they used to. Many studies claim, however, with considerable empirical evidence to show for it, that Americans have been sorting themselves out (the so-called Big Sort) through migration and switching identities into increasingly homogeneous electoral enclaves in which the minority party is becoming increasingly marginalized. Not only, then, is American politics increasingly polarized ideologically (and in terms of achieving any sort of policy consensus in Washington, DC), but the country itself is increasingly geographically polarized politically, albeit not at the level of sections or macroregions that characterized geographical polarization before the American Civil War but at the scale of counties, suburbs, and municipalities.

- The so-called Arab Awakening of December 2010 throughout 2011, with large-scale popular protests against long-established dictatorships across the Arab world from Morocco to Bahrain, has been widely interpreted in the outside world as a response to a growing disenchantment of youthful populations with increasingly superannuated dictators, high unemployment, the diffusion of satellite television news covering the Arab world, and the use of social media (such as Twitter and Facebook) to mobilize within and across countries. Such commonalities undoubtedly exist. But there are also very important differences between countries in their social and geographical characteristics that figure into how protests have played out and what has and what has not been achieved. For example, in Tunisia and Egypt relatively well-organized oppositional groups (including Islamists) based in the few major urban centers were able to overturn the longstanding rulers but without necessarily changing the governmental regime (meaning dominant interests or institutions such as the military) in place. At the same time, in Libya, a tribally organized society proved much more resistant to changing either personnel or regime, and in Yemen a geographically fractured (north/south and mountains/plains divisions) and extremely poor country is faced with the disintegration of any sort of central government rather than simply a change in either personnel or regime. Any adequate analysis of the Arab Awakening, therefore, must attend to more than generalizing from one case to the others or overemphasizing commonalities at the expense of a richer comparative and geographically based approach. This should also include consideration of how other countries may have subsequently intervened by backing or opposing the protests (such as France and Britain in Libya, and Saudi Arabia in Bahrain).

- Good climate and easily accessible water make Punjab the "breadbasket" of India. But the water table has dropped by some ten meters since 1973, and the rate of decline has picked up in both Indian and Pakistani sides of the region. Much the same story can be told for the northwestern Sahara aquifer shared by Algeria, Tunisia, and Libya. Each of these cases is an example of the classic "tragedy of the commons." Each farmer pumps at will, oblivious to the collective impact. Increasing scarcity encourages rather than undermines this. Everyone looks out for his own as access deteriorates. Powerful farmer organizations and underpricing encourage profligacy in use, particularly in agriculture. The problem is compounded when aquifers are shared by two or more countries. Coordination of usage is difficult enough within one country. Almost 96 percent of the earth's freshwater is stored as groundwater, and about half of this crosses international borders. This creates a major problem in geographical coordination. Yet it is encouraging to see many examples of successful agreements to manage cross-border water resources such as that signed by Brazil, Argentina, Uruguay, and Paraguay in summer 2010 governing extraction from the cross-border Guarani aquifer. Conflicts between neighboring countries over water resources may be inevitable, but their "management" by recourse to violence is not.
- Famines, hurricanes, and deadly heat waves have natural origins, but they are also intensely political with respect to the effects they produce. In the case of famines, it is often how food distribution is managed subsequent to local food shortages rather than the shortages per se that is crucial in determining the extent of starvation. With hurricanes the extent of damage depends as much on building standards and storm preparation as on the strength of the hurricane itself. In the case of heat waves, deaths are often seen as inevitable, particularly among the elderly and those vulnerable to high temperatures and humidity. Yet the extent to which heat waves produce otherwise unexpected spikes in deaths has as much to do with social and political organization in given areas as it does with the intensity of the heat wave itself. Studies comparing different neighborhoods of similar social composition during a heat wave but with differing degrees of social solidarity (kinship and family ties versus social isolation, etc.), access to air conditioning, robustness of the electricity grid under peak demand, and action by social-service agencies showed very different outcomes in terms of deaths. Even the seemingly most "natural" events, therefore, have different outcomes dependent on prior preparation and the political agendas this reflects.
- Naples in Italy has had a garbage "crisis" on and off since 2008. The promise to do something about it helped propel Silvio Berlusconi into

office as Italian prime minister in 2008. But the problem of disposing of the city's solid waste has never been solved. If one aspect of the problem is the overflow of local landfill sites and the lack of other means of disposal (such as incineration), the other is the inflow of waste to the region from other parts of Italy. Some of the inflow and the use of a refuse "crisis" to gain political leverage is undoubtedly the result of the working of local organized crime in Naples, the Camorra, which has turned waste "management" into one of their most profitable activities. Naples and its suburbs have become the "poster child" for a very modern problem: how to properly dispose of the waste of modern urban society when the problem is geographically extensive and the solutions have to be provided locally. So if socially it is an Italian problem, politically it is seen as one that Naples (and the surrounding region) must solve on its own. Most public goods (everything from highways and education to policing and fire protection) and, in this case, public "bads" (pollution of all sorts, poor building practices, lack of preparation for emergencies, etc.) are distributed unevenly. Naples has one very serious public bad. There is a more general principle at work here: some places have many good public services and few bads, other places have the reverse, often, as in this case, because of histories of economic exploitation, social deprivation, and related organized crime.

- The Eurozone was put together among a majority of the then-member states of the European Union in the late 1990s. The process of monetary unification, in which the euro replaced preexisting currencies among members, was avowedly a political act: designed to signify the ever deeper union that had inspired the founders of the European Union (as it had become) when it first started as a dream in the 1940s and 1950s. Unfortunately, the rules governing the fiscal behavior of the member countries were not rigorously based on centrally enforced norms, and there was no facility to create Eurozone bonds through which to establish the reputation of the zone as a whole within global bond markets. As a result, in the face of the worldwide financial collapse of 2007–2008, the weaker national economies of Europe were left without the ability to devalue their national currencies (now being part of the larger currency area), and the stronger national economies (particularly Germany) had to face the reality of bailing out the weaker members as the spreads on bond yields between their national bonds widened inexorably. The so-called peripheral economies of Western and Southern Europe within the Eurozone (Ireland, Greece, Portugal, Spain, Cyprus, and Italy) found themselves in a particularly disadvantaged position. A profound geographical fracture within the Eurozone, owing something in part to comparative histories of government profligacy, corruption, and tax evasion, but also to the weaker structural conditions of their economies when they entered the Eurozone in the first place (particularly a tendency to use devaluation as a tool of economic management and a proclivity for patronage politics that interfered with labor market, pension, and other reforms), lay at the heart of the crisis. The adoption of the euro without simultaneous European political union and fiscal integration seems to have prepared the ground for disaster. This was then enhanced by global investors' speculating on the odds of specific countries' successfully managing their economic weakness, as well as by the "verdicts" offered on their solvency by credit rating agencies, which in demanding more sacrifices from those countries not only hampered their economic growth but also eroded state fiscal sovereignty and undermined democratically elected governments. The crisis of the Eurozone and of the European Union is first and foremost, then, a geographical crisis between center and periphery in the absence of the sort of fiscal union that exists in the United States to manage divergent regional economies.

- On May 1, 2011, two U.S. Black Hawk helicopters took off from Jalalabad airfield, in Eastern Afghanistan, to penetrate Pakistan's sovereign airspace for about 120 miles to reach the hideout of Osama bin Laden in Abbottabad. They carried a team of U.S. Navy SEALs previously flown there from Virginia via Germany. Bin Laden's secret location in a small town not far from the capital of Pakistan had been pinpointed by CIA analysts after a satellite had captured the image of a white SUV belonging to one of bin Laden's couriers parked outside a large compound eight months earlier. This was the endgame of one of the most stunning events so far of the twenty-first century: the terrorist attacks of September 11, 2011, on the twin towers of the World Trade Center in New York City and the Pentagon in Washington, DC. Claimed as their handiwork by a shadowy network of Islamist terrorists based in Afghanistan, the attacks represent the most outstanding example of the increased importance of asymmetric warfare between powerful states, on the one hand, and fanatical groups expressing some sort of global or religious ideology who can now project themselves around the world because of modern technologies like jet planes and cell phones, on the other. The apparent leader of the 9/11 network, named al-Qaeda by its leaders, was a Saudi Arabian millionaire, Osama bin Laden, who had begun his career as a political militant working to remove the Soviet occupiers from Afghanistan in the 1980s. In the Cold War context of the 1980s, he had been an American ally against the Soviet enemy. Bin Laden was now rapidly named the world's most wanted man by the U.S. government. Although U.S. invasions of Afghanistan and Iraq in subsequent years were sometimes ascribed to the hunt for bin Laden and his network, it is now widely accepted that they became distractions from that pursuit. The two wars fit into an outdated mindset in the U.S. that remained locked into a state-to-state conception of world politics rather than seeing that bin Laden represented a novel phenomenon: the head of a network of militants with widely scattered sources of financial and intelligence support from elements in governing and military circles all over the Middle East and Southeast Asia (but particularly in Saudi Arabia and Pakistan). A major problem for the U.S. was that the governments of Saudi Arabia and Pakistan were both considered as "allies" when in fact they were anything but. Indeed, following the U.S. assault on Afghanistan in late 2001, bin Laden left his "safe haven" there for Pakistan, where he was to remain until May 2011. The record suggests that little energy was put into finding bin Laden until after the election of the Obama administration in 2008. Finally, in the raid deep inside Pakistan on the night of May 1, 2011, a team from U.S. Special Forces flew from Afghanistan to bin Laden's hideout in Abbottabad, Pakistan, a large house/compound, literally less than a mile from the entrance to Pakistan's most prestigious military academy. Bin Laden was killed in an assault on the house, and his body was later dumped into the Arabian Sea. The raid and killing were seen by a raptured audience in the White House in real time by satellite television from the scene. "Getting bin Laden," as the entire event was soon labeled, involved collecting intelligence surreptitiously within the territory of an "ally" (although now, it seems, hardly an exemplary one), violating that country's "territorial sovereignty" without even informing that country's government of what was afoot. At the same time, another neighboring country, Afghanistan, unwittingly provided the base for operations because it is still in the throes of a war in which the U.S. is a major protagonist that began in the aftermath of 9/11 with the presumed search for those responsible for the terrorist attacks. Bin Laden's body was disposed of at sea to prevent any possibility of a burial site on land becoming a pilgrimage site for disciples. Osama bin Laden may now be dead, but the war he helped to start still goes on. In this context, does it make sense to continue thinking of the world politically as simply a set of equivalently powerful sovereign states overlain on a map, which is the common sense of much of what goes for international relations thinking and journalism?

- With a death toll of seventy-seven, Anders Behring Breivik was charged in July 2011 with one of the worst acts of terrorism carried out by a single person in history. Initially ascribed to Muslim terrorists by myriad experts even before any facts were known, later reports revealed that Breivak, born and bred in Norway, had planted a bomb in a government building in the capital city, Oslo, and then gone to an island in a lake near the capital, where he proceeded to methodically kill young people there for a summer camp organized by the governing Norwegian Labor Party. Obviously, acts of this sort indicate someone with a crippled moral vision of how to relate to fellow humans akin to the genocidal maniacs of the Third Reich. Planned for many years, however, Breivik's own explanation of what he did, posted on the Web in a long statement detailing why he did it and how he hoped to recruit others to his cause, reveals someone deeply angry about what he sees as the religious-national "diluting" of Norway's population by recent waves of immigration, particularly that of Muslims. His ire was directed in particular at those Norwegians he saw as the instruments of this dilution through their commitment to "multiculturalism" and what he dubbed the "cultural Marxism" that in his mind underpins this. Drawing sustenance from anti-Islamic websites across the world, particularly in England and the United States, Breivik chose to frame his actions as the first steps in a crusade to reclaim Europe from what he, and the websites he drew from, see as an Islamic "invasion" of Europe. Rather than a narrowly Norwegian nationalist, therefore, identifying with a traditional nationalist agenda, Breivik can be plausibly characterized as a sort of a European "macronationalist" or a European mirror image of the jihadists of al-Qaeda, who have a territorially expansive vision of a world they too believe to have lost. Considerations of geographical scale relating to the visions and imaginations of their protagonists are vital to understanding this sort of thinking and the political movements around the world with which it is associated.

- Though organized violence by terrorist groups has been largely over in Northern Ireland since 1998, there are episodic outbreaks of riots in and around the boundaries of heavily segregated Catholic and Protestant neighborhoods in the largest city of Belfast and elsewhere. Much of the tension is in areas with relatively poor populations on either side of the religious divide (which correlates highly with either Catholic Irish Nationalist or Protestant British Loyalist proclivities). The fights are usually about defending communal identities by defining clear boundaries between adjacent groups. The scene of much intercommunal violence over thirty or so years of conflict, Belfast has seen its Protestants become a shrinking minority in some violence-prone areas such as the Ardoyne district in north Belfast, as the more affluent Protestants have moved out to Belfast's suburbs and socially mobile Catholics, usually with larger families, have moved in. The local Protestants face the prospect of losing their ability to hold the line between the two groups and thus perpetuate their own identity. Ethnic conflicts and struggles over political identities are closely bound up therefore with controlling territories and maintaining boundaries.

These are all examples of vitally important public issues that can benefit from a geographical framing. The framing not only allows for an integration of facts into a more meaningful story but also identifies how geographical factors can be crucial in thoroughly understanding a given political phenomenon—from global terrorism and ethnic conflicts to the politics of environmental degradation, water politics, the politics of famines and heat waves, the origins and spread of large-scale political protest, and the changing character of electoral politics. The trick in doing so is to think in terms of maps—maps that not only locate but also join together the places and regions inherent in the various stories (Foucher 2011). The map can be in the mind's eye, on paper, or on a computer screen. It is the "technology" or methodology most specific to geographical thinking of all genres, even if in recent years it has been less apparent in political geography than formerly. There is a map that brings together the facts of the killing and disposal of Osama bin Laden in May 2011. There are also maps that show who lives where in north Belfast, where public goods and public bads are to be found in Italy, how the Eurozone is structured and operates, how the Arab Awakening began and spread, and, if done correctly, how American party politics is polarized geographically. Of course, maps are constructed entities, so they are invariably selective and therefore must always be read with care.

The various public issues, however, can be examined in a wide variety of ways. Maps are invariably selective, as the U.S. red state/blue state example indicates. The map relationships we choose to emphasize must be selected on some basis. As we shall see, political geographers have done their selecting in widely varying ways reflecting very different theories about how the world works. The making of political geography is largely a story of how they have gone about doing so and how this has affected what they have decided to study.

The four main chapters cover the following ground. Chapter 2, "How Political Geography Is Made," introduces the making of political geography in four ways: in terms of how it has been made by authors with very particular relationships to concentrations of power, how the emphasis on different subject matter and different theoretical perspectives has evolved from the late nineteenth century to the present, how different meanings of the "political" have informed it down the years, and why political geography continues to matter in the early twenty-first century.

In chapter 3, "The Historic Canon," some of the classic works in political geography from the 1890s to World War II are given pride of place. We concentrate on some of the "big names" of the period and their influence, attempting to place both ideas and influence in the context of the time and the interimperial rivalry that characterized the period.

Chapter 4, "Reinventing Political Geography," examines the key themes and perspectives in a new political geography that began to emerge in the

1960s. This expanded beyond the traditional focus on national-states, their boundaries, and the global contexts in which they found themselves to a more general interest in boundary making as the process whereby power is expressed geographically at whatever geographical scale—local, regional, national, global—this occurs. From this point of view, political geography was redefined as the study of political practices producing boundaries and the impacts of these boundaries on the power and welfare of different social groups. Certain important theoretical statements and empirical case studies are used to highlight the new meaning given to political geography. Three "waves" of perspectives, representing distinctive philosophical and political positions, are seen as having swept successively across the intellectual landscape: spatial analysis beginning in the 1960s, political-economic perspectives in the 1970s, and postmodern perspectives in the late 1980s and 1990s. At the same time, the empirical scope of what was understood as political geography was also expanding from the geopolitics and spatiality (internal geography) of states (the classic concerns) to geographies of social and political movements, ethnic conflicts and nationalism, and place and the politics of identity. Each new "wave" seemingly brought with it new subjects of study. An attempt is made to tie the waves to the intellectual and political fluidity at the end of the Cold War, the huge social changes in Europe and North America in the 1960s and 1970s, and the recent emergence of a more globalized world in which established theories of politics and society seemingly offer less intellectual purchase than they once did.

Finally, chapter 5, "The Horizon," after providing an interpretation of the new post–Cold War geopolitical context that has emerged since 1990, identifies three sets of intellectual issues that beckon on the horizon of political geography and that are already becoming central themes of the field: the global politics of the physical environment, power and geographical scale in a globalizing world, and ethical choice and values in the geographical organization of politics. A number of empirical case studies and descriptions of particular books are used to provide examples of the type of analysis the new topics can elicit. We do not mean to suggest that other issues, such as those examined in chapter 4, are now of lesser importance, only that these three seem to offer the greatest challenge in the geopolitical context of the early twenty-first century. Political geography has never been anything if not a reflection of its times.

2

How Political Geography Is Made

In this chapter we want to offer four different angles or viewpoints on how political geography can be considered to have been "made," particularly over the past one hundred years. Each raises a general issue necessary to a full understanding of what follows in subsequent chapters:

1. We argue that knowledge is always situated in specific historical-intellectual contexts and power relationships that color both approaches and interpretations, but that *political geography* as a term meaningfully refers to the mediating effects of geography on politics.
2. We provide a brief history of the subfield and the varied perspectives and themes that have been adopted so as to orient the reader to the subject matter of subsequent chapters.

3. We survey the different meanings that have been given to the qualifier *political* and suggest that across most of these there is a definite distinctiveness to what is political and, hence, that a separate political geography makes sense.
4. We establish why the relationship between geography and politics matters in the contemporary world and, hence, why you should be interested in this book. We use six vignettes drawn from contemporary settings around the world to make some specific points about how the relationship of geography to politics still counts, perhaps more than ever.

POWER AND KNOWLEDGE

Political geographers, like many other social scientists, have become increasingly sensitive to the charge that they always look at the world from specific social and geographical "positions" such as American, white, male, gay, Catholic, and so on. Even as they endeavor to offer theoretical coherence and empirical evidence for the perspectives they bring to bear on particular situations, they cannot in good faith claim to be totally disinterested or impartial. In the world of human making there is no singular "view from nowhere" that can be invoked to justify this or that perspective as better than another. This does not mean that all knowledge claims are thereby equally valid. Seeing knowledge as produced in historical-geographical contexts does not entail a claim with which it is frequently confused: that all knowledge is simply *relative* to this or that historical-geographical context or social position. The point is that we should be alert to the contextual biases built into any and all knowledge claims (Agnew and Livingstone 2011).

At one time, however, a strongly naturalistic view of knowledge (that nature "speaks" to us unmediated by any sort of social or philosophical bias) did prevail almost totally, and the first generation of political geographers—such as Friedrich Ratzel in Germany and Halford Mackinder in England—used it to mask their profoundly nationalistic viewpoints. That said, they lived in an era when total devotion to one's nation was largely taken for granted by the upper-middle-class academics who wrote most political geography in Western Europe and the United States. The historical-geographical context matters, therefore, in understanding how political geography has been "made" down the years, just as it does for other academic fields.

More specifically, it is clear that upper-middle-class European and American men have long dominated the entire enterprise. Indeed, the rise of "intellectuals" as a class of people in Europe and North America is associated with the professionalization, commercialization, and legitimization of knowledge on the part of powerful institutions (states, universities, businesses) that were and largely still are dominated by white men from privileged social backgrounds. It is only within the past forty years that white men of other social classes, women, people from postcolonial settings (such as India and Africa), and other outsiders (including people from places such as Australia, China, and Japan) have made much inroads into the inner temple of academia in general and the study of political geography in particular.

This is important for two reasons. First, it means that until recently understandings of such key concepts in political geography as power, boundary making, and territoriality have reflected the understandings and practices of those long in command within the world's Great Powers such as France, Germany, Britain, and the United States. At one and the same time that they have claimed the mantle of "objective reality" and "science" for their accounts, they have also served different national interests with which they have explicitly aligned themselves. The growth of political geography elsewhere—for example, in countries such as Italy, Russia, and Japan—if not without its novelties and differences of emphasis, nevertheless tended to follow the intellectual tracks laid down by the currents of thought emanating from higher up in the global hierarchy of states.

The combination of claim to objectivity with serving a national interest has had a specific foundation: the implicit claim to offer a "view from nowhere." This is the idea that the whole world can be known totally from beyond the particular geographical and intellectual location of the person making the claim. But as the philosopher Thomas Nagel (1986, 25–28) has pointed out, there can never be complete objectivity; there is always an "incompleteness" to objective reality. In other words, objective reality does not exhaust reality. Rather, "any objective conception of reality must include an acknowledgment of its own incompleteness." The solution for Nagel is to "enrich the notion of objectivity" by adding the views of oneself and of other selves so as to incorporate the insights of subjectivity. As Nagel concludes, "to insist in every case that the most objective and detached account of a phenomenon is the correct one is likely to lead to reductive conclusions. I have argued that the seductive appeal of objective reality depends on a mistake. It is not a given. Reality is not just objective reality." So even though acknowledging the prospect of a knowing that is not completely subjective, Nagel is suggesting that the quest for complete objectivity is a fruitless one. There can never be a view from nowhere that is not also, and profoundly, a view from somewhere. The history of political geography certainly provides support for Nagel's argument.

Widening the social origins of those studying the field, however, opens up the potential range of geographical and intellectual locations from which the problems of the field can be studied. Indeed, it also opens up the possibility of redefining the kinds of problems that the field studies. Moving away, for example, from an exclusive focus on "states" as if they have always been and are still the single fount of power in everyday life, we can expect to see a distinctively different set of "subjectivities" entering into the field, thereby "enriching" objective reality in new ways, beyond that of the national interests that once reigned supreme—given the lower attachment of many new recruits to this or that national interest, particularly of the Great Powers of the day, the increased internationalization of academia as people move between universities in different countries, and the greater variety of universities and educational experiences involved in the enterprise.

The second reason why wider recruitment into the ranks of political geographers matters in terms of the sort of knowledge it produces is that political geography has long presented itself as a problem-solving activity, to be called on by those in seats of power to help resolve their dilemmas. This technocratic impetus was there from the start. A technical or control impulse characterized the natural and social sciences from their beginnings as university disciplines in the late nineteenth century (Stanley 1978). But it was never without challenge. Within late nineteenth-century Europe, for example, radical geographical thinkers such as Peter Kropotkin and Elisée Reclus provided the basis for a critical approach to political geography. But this proved out of touch with the times, a period of intense national and interimperial rivalry, and unattractive to those intent on professionalizing geography and other fields such as political science within the expanding university systems of the Great Powers. Attracting government and business support for the new fields required providing such institutions with problem-solving capacity, not criticizing their activities and wherewithal. Only recently have "minor figures" such as Kropotkin and Reclus been rediscovered and accorded serious attention (e.g., Clout 2005), notwithstanding a tendency to laud their intellectual independence and political radicalism without much critical examination of some questionable ideas (see, e.g., D. Miller 1986, on the problems with Kropotkin's social analysis).

It is the rise of critical perspectives in social theory, initially associated with the sociology of knowledge and some varieties of Western Marxism, but later expressed most forcibly in feminism and some types of poststructuralist philosophy, that have led away somewhat from the problem-solving focus on states and hagiographic accounts of the "founders." In particular, feminist and other thinkers have opened up for discussion such issues as reason versus affectivity (emotion) in human action and notions of a homogeneous (presumably masculine) national public upon which much discourse in political theory has long relied (e.g., I. M. Young 1987; Massey 1994; Kofman 2008). Problem solving involves the ability to specify whose problems are given pride of place but also, much more importantly, how those problems are defined and addressed. Feminist thinkers, therefore, not always but largely women, have questioned the idea of problem solving as an academic focus without prior attention being given to the ways in which it is done and for whom.

What was lacking until recently has been an appreciation of the relationship between knowledge and power. We have learned from such writers as the French philosopher Michel Foucault (e.g., 1980) and the Palestinian American literary theorist Edward Said (e.g., 1978) that knowledge consists of "discourses" or sets of ideas, terms, and connecting phrases that arise in distinct historical-geographical contexts (such as Vienna in the 1890s, Paris in the 1970s, or Siena in 2010) and that persist because they are adopted by others and become part of a "common sense" that defines a discipline or field of study (such as political geography). This is not to say that all knowledge production can be read off from the social and geographical backgrounds of its producers. Since the European Enlightenment of the seventeenth and eighteenth centuries, there have always been currents of thought that challenge and threaten dominant conceptions of knowing. For a much longer time there have always been "outsiders" critical of the intellectual status quo. One feature of modern intellectual history, therefore, has been an increasing "suspicion" of established knowledge claims and the social interests such claims uphold. The social construction of knowledge in particular historical-geographical contexts is not the same as social determination *tout court*. From this point of view, indeed, some claims are better than others, and there can be progress in political geography (Bassett 1999).

Discourses are both enabling and disabling: they allow us to construct research projects, for example, but they also direct our minds in certain directions rather than in others. Within the Anglo-American social sciences in general, for example, particularly since World War II, thinking with geographical concepts has not had widespread support. At the same time, until the 1960s, political geography was largely devoted to examining the creation of the world political map and the emergence of Great Power spheres of influence without much if any attention to its political-theoretical assumptions. This was the dominant discourse, reflecting the times in which it arose: the new world of states at the end of European colonialism and the day-to-day operation of the Cold War. It is only with the recruitment of people into the field from a wider range of backgrounds, the emergence of a globalizing world economy, and the ending of the Cold War that new perspectives have found acceptance. The very political and social instability of the current world encourages the sort of intellectual experimentation that has characterized political geography in the recent past.

To reiterate, this is not to say that authorship and intellectual innovation are totally determined by the time and place in which an author is situated.

Figures like Kropotkin and Reclus strongly suggest the limits of contextual explanation. It is more that contexts tend to impose limits on what can be thought and written. Even many late nineteenth-century anarchists and socialists found colonialism not only acceptable but also praiseworthy for the possibility of "progress" that it brought to regions they thought of as "backward" and benighted. Today, Western intellectuals tend to reject the biological basis to races, whereas many of their early twentieth-century counterparts found this idea not only acceptable but absolutely fundamental to their understanding of the world. In a somewhat different vein, Islamic disputes over the treatment of women pose difficulties for those in Europe and North America who believe that Western capitalism or statehood are the singular sources of oppression in the world. In the face of global environmental crises, it has become popular in some circles to argue for a less human-centered understanding of economic development, incorporating other species into calculations about economic development and its effects. These are all examples of how historical-geographical contexts condition the possibility of different kinds of knowledge seeking and limit exploration of the ideas that can be used in that task.

Despite the history of contending perspectives and discursive "breaks" because of contextual influences, there is a degree of coherence to the term *political geography* that has persisted down the years and from place to place. This coherence has three dimensions to it. One is a persisting focus on a common set of concepts—particularly, boundary, territory, state, nation, sphere of influence, and place—even as these concepts have been endowed with changing meanings and applied in different ways (for example, within rather than just between states). One can plausibly claim that interest in and use of these concepts predates the adoption of the term *political geography* itself. Hence, it is the concepts and not the label that packages them together that constitutes the field as such. Certainly, such luminaries in the history of political thought as Aristotle, Sun Tzu, Machiavelli, Hobbes, Montesquieu, Madison, Herder, Rousseau, Hegel, Marx, and Gramsci addressed questions of statehood, citizenship, and the geographical distribution of power. They can be thought of as proto–political geographers (among other things, of course). But there is nevertheless the sense that since the late nineteenth century an academic division of labor has evolved in which a niche called "political geography" has emerged and which has become institutionalized in university courses, academic organizations, conferences, and academic journals.

The second dimension is a theoretical focus on trying to discover the ways in which geography (defined in various ways, from the physical geography of the world to the distribution of ethnic groups and economic resources) mediates between people, on the one hand, and political organization, on the other. There is a persisting tendency to insist that politics cannot be adequately understood without understanding the geographical contexts in which it takes place, from global geopolitics at one end of the scale to local politics at the other. What has been understood as "the geographical" has undoubtedly changed down the years. At one time it was understood almost entirely in physical terms, whereas today the sense is almost entirely of the human organization of the earth's surface. Notwithstanding these important differences, however, it is clear that a common commitment to placing politics in a geographical framework has remained a constant.

Finally, there is a sociology to academic subfields manifested in professional organizations (such as the Political Geography Specialty Group of the Association of American Geographers or the Political Geography Committee of the International Political Science Association) and journals (such as *Political Geography* and *Geopolitics*) devoted to their subject matter. This reflects the way in which university graduate training is organized to induct students into "areas of knowledge" by preparing them to work, publish, and teach in a given subarea. In this understanding, political geographers constitute a sort of "intellectual tribe," sharing certain norms of academic practice—how research is done, how articles are written, whom the articles are written for—that differ from those in such adjacent areas as international relations, cultural geography, and economic geography. Given the relatively marginal status of geography as an academic field in higher education, particularly in American academia, the "disciplining" of students into the field is possibly less problematic than in fields that are more dominant academically and in which single intellectual orthodoxies often hold sway. Nevertheless, there is always the danger that in encouraging disciplinary norms at the expense of alternative viewpoints intellectual innovation is restricted. Yet there is little if any evidence of this in contemporary political geography. Its heterodox character suggests that "disciplining" has not worked too well. Much of this book is taken up with trying to organize and describe the contemporary field's variety rather than the inculcation of dominant "truths" that seems to be all too characteristic of so many fields.

Nevertheless, even when theoretical perspectives differ and writers use different terms to label what they do, a coherent field of political geography can be defined. Increasingly, disciplinary labels are seen as problematic. Indeed, you may be using this book in a course with a quite different title such as "Place and Politics" or "Geopolitics." Be that as it may, if you are concerned with the intersection of geography and politics, you are doing political geography or some aspect of it!

THE HISTORY AND LANGUAGE OF POLITICAL GEOGRAPHY

The term *political geography* in contemporary usage, like so much of the terminology and labels in the social sciences, dates from late nineteenth-century Germany. Arguably, the term was first coined in print by the French *philosophe* Turgot in 1750. This is the beginning of the period in European intellectual history when the modern social sciences began to emerge and when "the study of the underlying forces that shape our will was substituted for the idea of human agency that lay at the heart of the now superseded political sciences" (Wokler 1987, 327). The older idea of politics as shaping the world, explicit in the Lorenzetti frescoes discussed in the preface, was now replaced by the new idea of politics being shaped by other forces such as the environment or the economy. But it is to the German geographer Friedrich Ratzel in 1885, in an article on the "new political map of Africa," that the history of the subarea of academic geography known as political geography is usually now traced. Ratzel published the first book with political geography in its title (*Politische Geographie*) in 1897. Only recently unified politically, Germany was the center of attempts at rationalizing social knowledge in the interests of state development through the establishment and state support of universities as research rather than as simply teaching institutions. Other Great Powers, such as England, France, and the United States, rapidly followed suit. In the United States, universities such as Johns Hopkins University in Baltimore and the University of Chicago were directly modeled on the German prototype research university with an emphasis on narrow specialization and the training of postgraduate students as future researchers.

As the study of and writing about the earth as a whole, geography is an old field, but it was professionalized within the university relatively late. Only in Germany were there well-established chairs (or professorships) in geography in the mid-nineteenth century (Sandner 1994). England, the United States, and France acquired their first geography departments only in the last two decades of the nineteenth century and first decade of the twentieth. Consequently, the identification and practice of subfields also came late. In the field of geography this was complicated by the fact that many of its proponents tended to see it as an "integrative" field and resisted breaking it down into specialties, even if this seemed necessary for the purposes of research. In France, for example, Vidal de la Blache resisted the idea of separate geographies, each with their own qualifier. For him there was one geography or none at all. Though there was a long-established tradition in German-speaking Europe of "state geography," as opposed to a "regional geography" based on natural divisions, devoted to showing how social as well as natural features were bound up with but often contradictory to existing geographical patterns of statehood, its naturalizing of the state was incomplete. Only with Ratzel does "the state take possession of geography and become its supreme object." (Farinelli 2001, 44). In the subdividing of geography, Ratzel was the first to identify a separate political geography. Even as he did so, however, Ratzel was championing the geography of the state as the centerpiece of geography as a whole. In this intellectual world, the political was not a separate sphere but the central one associated completely with statehood. Ratzel is thus a theorist for his time. One in which a new "aristocratic-bourgeois state," to use Farinelli's term, was in the ascendancy in the organization of knowledge in general, not just in relation to self-confessed geography. Of course, this was also the time of a reinvigorated European colonialism, related in large part to German attempts at "catching up" with Britain and France in empire building following the recent unification of Germany and the defeat of France in the Franco-Prussian War of 1870.

The space or area occupied by the state and its position on the world map were seen by Ratzel as the key concepts of his 1897 book. Rooted in a particular space, a state expresses both a material relationship to its space through the "soil" (an idea that the German Chancellor Bismarck used freely to argue for German unification) and a spiritual relationship through the peculiarities of the occupying national group. States can only thrive, however, if they expand into other territories to express their vitality and their higher "cultural level." To Ratzel, the size of a state is one of the measures of its cultural level. Whether war is inevitably the mechanism of expansion is not made clear. Nevertheless, Ratzel's entire approach is premised on the idea that a state's position or location in relation to other states makes it more or less vulnerable to the expansion of other states. A state such as Germany is thus more vulnerable than an England because it is surrounded on all sides by other states. The combination of German vitality and geographical vulnerability produces the need to expand.

Though Ratzel and like-minded thinkers saw their writing in "scientific" terms, as reflecting natural processes or laws that lay beyond human command, they also saw their work as serving the goals of the particular states in which they lived. Thus for Ratzel the need was to offer a "scientific" account of why Germany required a larger space than it currently occupied. To Halford Mackinder, in England, the need was to use a global model pointing out certain "natural facts" about the distribution of the continents and oceans to represent the threat from Germany or Russia in the heartland of Eurasia to the vulnerable British Empire scattered largely around its fringe. This fusion of a claim to scientific objectivity with particular national purpose was to characterize political geography throughout the period of intense interimperial rivalry between Britain, France, and the United States on the one side and the rising powers of Germany, Japan, and Italy on the other. The geopolitical ideas of Ratzel and Mackinder were of greater appeal to the latter than to the former because they provided something of a broad guide to action for those attempting to challenge the geopolitical status quo.

The term *geopolitics* first arose in this context. Invented by the Swedish political scientist Rudolf Kjellén in 1899, geopolitics referred to the harnessing of geographical knowledge to further the aims of specific national states. If Kjellén was concerned to dispute the claim of Norwegian nationalists (Norway was part of Sweden until 1905) that the mountain spine down Scandinavia constituted a natural boundary between two distinctive peoples by arguing that seas and rivers were much more significant, the term *geopolitics* came to be applied by German thinkers in the 1920s and 1930s, most notoriously Karl Haushofer, to formal models of Great Power enmities based on their relative global location and need to establish territorial spheres of influence to feed their urge to expand. Through this formalization, Ratzel's idea of states as organic entities came to inform, if hardly to direct, German foreign policy after the Nazi accession to governmental power in Germany in 1933. After World War II the association with the Nazis gave the word *geopolitics* a negative connotation. Though used informally to refer to the geographical structure of international relations in the 1950s and 1960s, it has only been since the 1970s that the word has reentered political geography. Today it is typically used to refer to the ways in which foreign-policy elites and mass publics construct geographical images of the world and use these to inform world politics. But there are those, often historians and journalists more than professional geographers, who use the term either to invoke old-style environmental determinist positions (such as Mackinder's) or to introduce the direct effects of ad hoc "geographical factors" into their stories about the feats of heroic leaders and the dilemmas facing their states in different parts of the world (e.g., Black 2009, Kaplan 2010).

There was a current of thinking, largely developed in the United States after World War I, but also present in France and elsewhere, that took a more pragmatic view of the purpose of political geography than that of German *Geopolitik*. In this perspective, the need was to identify political-geographic problems (such as colonialism, the protection of minorities, international cooperation, raw materials, and national boundaries) and carefully detail their worldwide incidence and significance for world politics. The American Isaiah Bowman was the great advocate of this task for political geography. This was a form of state geography in which few if any questions were asked about statehood per se (in this sense, Ratzel is a more interesting thinker, whatever one thinks of his ideas) but much attention was paid to the pressing practical problems of the time in a frame of reference drawn largely from U.S. President Woodrow Wilson's outlook on national self-determination and the need for international collective-security institutions such as the League of Nations. If in later years Bowman liked to distinguish political geography from geopolitics, it is more because he saw political geography as a state geography than as geography of the state. Afraid of the double meaning of "political" in political geography—it could mean a political perspective from a particular standpoint as well as an enumeration of the conditions under which politics is practiced—Bowman hewed to the second so as to better portray his work as nonpartisan and scientific.

Bowman's fear that geopolitics would taint political geography, however, proved well founded. Though American political geographers like Bowman, Richard Hartshorne, and others were actively involved in providing the U.S. war effort during World War II with their brand of geographical synthesis, political geography emerged from the war years doubly tainted. As the U.S. government began its very own brand of geopolitical reasoning with the onset of the Cold War with the Soviet Union, open talk of geopolitics was associated with the worst excesses of Nazi expansionism. This served to ghettoize political geographers, who had nothing to say in response, even as diplomatic historians and political scientists invented their very own geopolitics without the word (based on such ideas as the *containment* of the Soviet Union, the *domino effect* connecting events at a distance from the United States to the homeland, and *national security* to identify an interest beyond that of any particular American). It also served to encourage a retreat by geographers in general from engagement with political issues, even of the relatively benign form engaged in by Isaiah Bowman. Though courses in political geography continued to be taught in the United States and elsewhere and textbooks modeled largely on Bowman's ideas continued to be published, little or anything that could be described as empirical research or critical thinking about the geography of the state or politics more generally appeared during these years. Certainly in the United States, and elsewhere in the English-speaking world, political geography became the "wayward child" that the cultural geographer Carl Sauer said it was in the 1950s and the "moribund backwater" that the economic geographer Brian Berry described it as in the early 1960s. In a sense, of course, a kind of unexamined but influential "political geography" did live on—in the practices of foreign-policy elites in the United States and around the world.

The exception who helps to prove the rule was Jean Gottmann, a French-trained geographer of Ukrainian-Jewish background. At the time, however, his work had little if any influence, at least in the English-speaking world. His brilliantly perceptive 1952 book, *La politique des Etats et leur géographie*, is based on a model of state development that Gottmann was to develop throughout his later career. He saw the political partitioning of the world as the result of the interaction between forces of external change (*circulation*) that move people, goods, ideas, and information and the territorially based beliefs and symbols (*iconographies*) that build group identities and create communities. The relative balance between these sets of forces determines the degree to which a "system" of states is open or closed. This historical approach to political geography was out of favor at the time because of the overwhelming emphasis given by dominant figures such as Hartshorne and Stephen Jones to the seeming perpetual functions performed by international borders and the obsession with distinguishing what was "geographic" from what was "historical," presumably studied by nongeographers. In 1961 Gottmann published his magnum opus, *Megalopolis,* a book in which he implied that the balance was shifting historically from national states to city networks as the basis for political geography (see chapter 4).

Gottmann long remained an iconoclastic figure. Without a permanent university appointment in either France or North America for many years, he engaged in a long period of movement to and fro across the Atlantic. His very mobility and life experience probably explain the critical approach he brought to his political geography. Unlike so many political geographers, he was not a prisoner of national horizons. His exposure to social and political theory also contributed to a much richer understanding of political-geographical concepts and analysis than that of any previous self-confessed political geographer. To him the political was not just a reflection of the environmental or the economic but had an autonomy and a primacy in its own right.

It is fair to say that Jean Gottmann was modern political geography's first real intellectual. Yet by the late 1960s he was no longer alone as a theorist of political geography. The period 1965–1970 marks the beginning of a serious revival of interest in political geography as more than a descriptive recital of particularities about boundaries and state characteristics. The revival initially had two elements: one was a focus on using electoral data and electoral districting to explore claims about how space mattered to political behavior (particularly the so-called neighborhood effect or impact of local voting traditions on electoral choice); the other was an interest in the historical geography of European state formation. The participants came from a variety of academic backgrounds and found common cause in such forums as the Committee on Political Geography of the International Political Science Association.

By the early 1970s, figures such as Kevin R. Cox, David Reynolds, and Richard Morrill in the United States, Ron Johnston in New Zealand (and later England), Peter Taylor in England, Paul Claval in France, Stein Rokkan in Norway, and Jean Gottmann (finally settled at Oxford University in England) were establishing a "new" political geography by distancing themselves from much of what had come before and seeing their endeavor as integrating political geography into the "mainstream" of social science. Much of this new political geography was modeled closely after the "spatial analysis" that swept through geography in the English-speaking world in the 1960s. It was based on searching for geographical patterns and theorizing about their origins (O'Loughlin 2003). Much of the theorizing drew from social-psychological thinking about political intentions (as in voting studies) or the comparative sociology of state formation (as in Rokkan's work on Europe). What was particularly striking, however, was the departure from a strict focus on national states to consider geographies of boundary making within states, electoral geographies (although the French tradition of electoral sociology had taken an ecological approach to election results since the early 1900s, it had, at its origins, tended to correlate outcomes with unchanging local features such as geology and soils), and geographies of conflict between social groups. It can be seen, therefore, both as a reaction to the political and social upheavals of the 1960s and the need to address these in a manner acceptable to current canons of scholarship.

This was not to last in dominating political geography. The development of a "radical" geography in response to both the Vietnam War and the civil rights struggles of the late 1960s produced a critique of spatial analysis as politically conservative, theoretically limited, and Eurocentric. In its place came "political-economic" perspectives drawing on Karl Marx and a range of other thinkers (particularly the American sociologist Immanuel Wallerstein) and claiming to find the roots of political territoriality in the history of the capitalist world economy and global competition between the Great Powers. Not only figures associated with radical geography, such as Yves Lacoste, Richard Peet, David Harvey, Doreen Massey, and Neil Smith, became influential, but also established figures in political geography, such as Peter Taylor and Kevin Cox, adopted various political-economic perspectives and put them to work in their research. What they shared in common was a critical view of contemporary world politics and society as unequal and hierarchical, a desire to communicate with critical social theorists in other fields about the importance of political-geographical mechanisms such as boundaries and core-periphery relationships in creating and maintaining the structures of inequality, and a commitment to political activism, often in the classroom rather than extending into the world at large. What remained more problematic was the status of the political in the political-economic perspectives, as most of the important figures associated with the perspective tended to diminish it to the advantage of the economic. If classical political geography saw the physical environment as driving the political, the political-economic perspectives tended to put the economy (and its agents in business/capital) in the driver's seat (Taylor 2003).

By the late 1980s the sense of philosophical certainty and the particular political commitments associated with the political-economic perspectives met with increased resistance. This was partly a product of more conservative times in countries such as Britain and the United States where post-1960s political geography was best rooted, but also of increasing skepticism about the possibilities of top-down political change and "all-knowing" theorists telling people what "really" matters and what they should do. A diffuse but nonetheless important reaction set in. Plausibly labeled *postmodern*, to signify perspectives that reject the modernist enterprise of social theory without knowing subjects and the idea of unsituated objective knowledge, this wave has had both direct and indirect effects in political geography (Sharp 2003). On the one hand, it has stimulated research attuned to the experiences of diverse groups (as, for example, in Paul Routledge's [1992] research on Indian villagers protesting against a government military facility) and the languages in which political disputes take place (as in Gearóid Ó Tuathail's [1993] examination of American discourse over the Gulf War). On the other, it has encouraged more modernist researchers to take the role of discourse and language more seriously, in rhetorical and communicative as well as constitutive roles. This is apparent, for example, in Agnew and Corbridge's (1995) book on the history of geopolitics as put into practice by political elites. Whatever these differences, writers within this broad perspective do tend to give "the political" a primacy that it has not hitherto had in the history of political geography. In this respect they see the other perspectives based as they are in environmental and economic determinisms of various sorts as representing the "antipolitics" that has been deployed against "politics," in the active sense of the word, throughout the long history of political philosophy (Howard 2010).

Contemporary political geography, therefore, is a mix of all three of the waves that have washed over the subfield since the 1960s. This composes an infinitely more complex and nuanced picture than in previous eras when dominant perspectives and personalities were easy to identify. In turn, this pluralization reflects the vast expansion of universities and the incentives thus provided to theoretical novelty and publication to indicate productivity by such enterprises as the Research Assessment Exercise in Britain and other competitions for university funding by governments to establish "new" approaches and distance these from established ones. But it is also due to the intellectual vitality of political geography, which has passed from a singular focus on statehood and formal geopolitical models as timeless truths to a wider range of subject matter concerning the politics of territory and boundary making under distinctive historical and geographical conditions.

THE MEANING OF "POLITICAL"

Politics is about struggles for power to exercise control over others and self, satisfy interests, and express or gain recognition for identities. That much is widely accepted, if with wide differences over whether to emphasize control, interests, or identities. What is less so is how the political originates and how important it is in relation to other "dimensions" of human life such as the economic or the cultural. Until recently, political geographers seemed rather agnostic and casual about the nature of the political (Painter

2008). They have tended to see states as the singular source and focus of power and to see power as the ability to coerce others, but without much if any explicit discussion of where they might stand beyond this bland acceptance of conventional wisdom among journalists and many politicians. Few political geography textbooks pay any attention to the meaning of the political and proceed seemingly oblivious of the need to critically examine the very qualifier they insist on putting in front of *geography*.

In practice, *statist* or *liberal* conceptions have prevailed. The former sees the (national) state as the singular source of identities and interests with persons as the agents of the collective enterprise represented by the state. Politics is a deadly serious business in which the political is the arena of authority in which absolute decisions are made and control is exercised. If the early-modern English political theorist Thomas Hobbes is one of the authors of this perspective, its clearest expression in the twentieth century appears in the writing of the German legal theorist Carl Schmitt. In his view, the essence of the political lies in distinguishing "friends" (those associated with a particular "form of existence") from "enemies" (adversaries of this way of life who would negate it given the chance). Both totalitarian, as in Nazi Germany and Stalinist Russia, and national-security politics, such as that on either side during the Cold War, rested on this meaning of the political. But one can recognize elements of it in the geopolitics emanating from the works of Ratzel and Mackinder. A potentially more benign version of this would be the civic nationalist or patriotic claim of "love of one's country" as the essence of the political, requiring as it does a clear sense of who is inside and who is outside the common project (Viroli 1995).

A logically similar, if politically more pluralistic, conception of the political is found among so-called *communitarians*. From their point of view, politics is about associating with others to express identities and pursuing the common interests that such identities then define. In this case it is the sociopolitical group (and its expression in distinctive institutions) rather than a state as such that is usually the object of affection, meaning, and belonging. Much of the literature on multiculturalism in contemporary social science rests on this meaning given to the political, notwithstanding the "liberal" politics (usually involving government assistance to this or that group) that often accompanies it (e.g., I. M. Young 1990).

The main target of both the statist and communitarian perspectives down the years has been the liberal one. In this understanding, the political should never be about control or identity but rather about procedures for discussion and compromise between the distinctive interests of different individual persons and the factions in which they coalesce. Associated classically with such theorists as John Locke and James Madison, the American philosopher John Rawls (1971) is perhaps its most distinguished contemporary exponent. Interests are seen as emanating from society, and hence the state has come into existence to manage and adjudicate between private interests. Politics, therefore, is necessary but should not become too "serious," otherwise it undermines the ability to manage divisions within society. It is about who gets what and where but, above all, how. The liberal perspective, however, has fragmented. On the one side are those who stress the primacy of individual economic motives and interests in politics, and on the other are those, such as some liberal feminists, who stress the pluralism of socially created differences that empower some groups (such as men) more than others. Spatial analysis in political geography tends to rest on liberal assumptions about both statehood and the primacy of individual economic motives.

Increasingly, however, the "classical" views have been challenged from two different directions. One perspective, associated with the political-economic critique that became important in political geography in the 1970s, tends to conceive of the political as supplementary or complementary to the economic, which, in the "last instance," is seen as determining of the nature of modern capitalist society. The dominance of capitalists as a class within states and in conflicts between states for capital accumulation at a global scale is seen as making the world go round. Politics is seen as the price that has to be paid for bringing other classes into "public life" and thereby legitimizing national statehood and the international contests such statehood engenders. Though frequently casting the political as simply "functional" for the economic, in the hands of some theorists (such as the English sociologist Michael Mann) this perspective has opened up to consider the relative autonomy of statehood from singular class imperatives and for the emergence of bureaucratic groups with their own identities and interests. Two giants in the history of social thought now appear as engaged rather than alienated: Max Weber, the sociologist of bureaucracy, complements Karl Marx, the economic theorist of capital accumulation.

The second direction is more radical in departing from both the classical views and the political-economic critique. Its major difference lies in seeing power as enabling as well as coercive (e.g., Allen 2002, 2003). From this point of view, the political is the means or modality for agency: the ability to act, resist, cooperate, and assent, as well as the ability to control, dominate, co-opt, seduce, and resent. In this understanding the political is no longer reducible to the economic, coercive activities by some over others, or states and their conflicts. Rather, the political exists whenever power is exercised in struggles over collective goods and identities, including the language that is used and the experiences or examples that are given a privileged position, reflecting the intellectual dominance of Europeans, men, white women, or whomever. From this point of view, articulated most forcibly in Joan Scott's (1992, 37) argument about the need to understand "experience" as always preinterpreted by the terms used to express it, any kind of social research "entails focusing on processes of identity production, insisting on the discursive nature of 'experience' and on the politics of its construction. Experience is at once always already an interpretation *and* in need of interpretation. What counts as experience is neither self-evident nor straightforward; it is always contested, always therefore political."

At first sight, this might seem like expanding the concept of the political almost to the point of meaninglessness. When everything is political, nothing is political. But it can be rescued in any of three ways, drawing in various ways on such twentieth-century political/social theorists as Antonio Gramsci, Hannah Arendt, Michel Foucault, Gilles Deleuze, and Bruno Latour. One is by emphasizing that historical configurations of power give rise to *hegemonies* (mixes of coercion and consent) exercised by dominant social groups or states (depending on the context). The widespread availability of power, therefore, never guarantees its equal distribution. Historical bias is built into its geographical distribution from place to place (see, e.g., Agnew and Corbridge 1995). A second solution is to envisage the world as engaged only indirectly through the *discourses* that provide the logic and language for practical reasoning. Such discourses are by definition hegemonic, they provide the direction and meaning to life, but they can also be challenged insofar as the magic of their words is identified and exposed as biased in favor of this or that state or social group. It is this larger definition of power, and particularly the concern for the operation of power through public discourse and in the action of different groups in relation to hegemonic projects, that characterizes the postmodern perspectives that became important in 1980s political geography (see, e.g., Ó Tuathail 1996). The term *postmodern* is not without ambiguity. What is most at stake, however, is the idea that the very process of defining the terms in which the political is addressed is itself political (Butler 1992, 7; also J. W. Scott 1992). Third, and finally, power can be thought of as implicit in the practices of *actor-networks* that connect, entrain, and shape all social activities. In so-called actor-network theory, power is the resource networks provide to actors (human, animal, and technological) to make networks serving business, associational, and political purposes work (see, e.g., Thrift 2000). The relational practices and performances that make up political networks, therefore, constitute the political in this solution to the narrowing of the political from the realm of everywhere to something more specific. Whether we might not well be back here with a substantively liberal understanding of the political is open to question.

DOES POLITICAL GEOGRAPHY STILL MATTER?

The world is undoubtedly a "smaller" place than it used to be, in the sense of the relative ease with which information, money, goods, and people can now whiz around the world. Although few places on the earth's surface have been totally out of touch with those elsewhere since the fateful expansion of Europeans into the rest of the world beginning in the late 1400s, the interlocking of places today through economic and cultural transactions is both more intensive and more important in people's everyday lives. In the world of the Internet, satellite television, global production chains, and interconnected world financial markets, the "old world" of contending empires, nationalist politics, and so on is often portrayed as about to fade into the sunset. But does the "time-space compression" of the world about which we now hear so much also signify the imminent collapse of the political boundaries that still mark the world map, decreased hostility between social groups over the spoils of political and military dominance, the increase in equality between places in the global distribution of power, and the demise of territorially based political identities? Or is it more that the world's political geography is currently experiencing a set of stresses from economic and technological change that affect the meaning and roles of territory, boundaries, spheres of influence, ethnic geographies, and place-based politics? In this historical context, which conception of the meaning of "the political" outlined earlier makes most sense to you? Are the classical statist and liberal views losing their theoretical grip? Bear these questions in mind throughout the book.

In the rest of this chapter, six vignettes or stories are used to explore some issues that are desperately important in the contemporary world and that all have vitally important political-geographical elements to them. They are all types of issues that are constantly in the newspapers, even if the particular situations chosen to illustrate the general points they make are possibly idiosyncratic or exotic to the reader. Our purpose here is to show how situations that you may not have thought about previously as political-geographical ones can be thought of fruitfully in this way. Far from dying out in the Internet world, questions of political accessibility, boundary making, and self-expression and identity creation through territorial control are very much alive and kicking. These examples are all developed at greater depth than those provided in chapter 1. The focus here is much more analytic than simply suggestive.

What is clear across all of the vignettes is that political-geographical considerations still matter enormously in people's lives, if often in different ways than in the past. Take, for example, one of the major problems of contemporary American society and the subject of the first vignette, the issue of drug addiction and the illegal trafficking that services it. One can argue that legally available alcohol, tobacco, and prescription drugs are all harmful when abused and that the same treatment for currently illegal stimulants might lead to a more "rational" approach to reducing addiction and its largely negative social effects. Be that as it may, official policy in the

United States since the Nixon presidency of 1969–1974 has been to pursue a "war on drugs" through both domestic advertising campaigns against the dangers of drug abuse and attempts at reducing the importation of drugs into the country. Reducing access to illegal drugs such as heroin and cocaine through reducing the supply and raising prices is therefore one of the main features of U.S. policy. This relies on the idea that if only supplies can be stopped before they arrive in the United States, then drug use can be curtailed. The boundaries of the United States thus become the "front line" in the war against drugs. But the drug flows themselves are hardly structured on a country-by-country basis. They form global webs that increasingly challenge the idea of a political world neatly divided by state borders. Efforts at curtailing them likewise rest on the cooperation and sometimes the cross-border intervention of police and military forces from destination/consumer countries such as the United States in producing ones such as Colombia or Mexico. There is a global, not simply an international, dimension to the world drug business and attempts to interfere with it.

The drug problem is an immense one and an important feature of the globalizing world economy, missed by those who focus relentlessly on national statistics of legal trade to argue that the world economy is still an international rather than a globalizing one. The global drug business has a distinctive geography to it with the major flows of heroin coming mainly from Afghanistan and Southeast Asia, cocaine from Andean South America, and methamphetamines from a variety of places. All flows lead to the major consumer countries in Europe and North America, where the majority of addicts and recreational users are located. One interesting trend of recent years is that the narcotics business is increasingly localized as homegrown marijuana and methamphetamines substitute for heroin and cocaine, and supplies of the latter tend now to come from shortened supply chains compared to twenty years ago. North America remains the largest single cocaine market with about 40 percent of global cocaine demand, and it also gets most of the profits generated by drug trafficking (UNODC 2010). For example, only 3 percent of the value of U.S. cocaine trade goes to the farmers and traffickers in the Andean countries, while 70 percent of profits go to the U.S. mid-level dealers and consumers. In 2009 Afghanistan was the source of 89 percent of illicit opiates, destined mostly to Western Europe and Russia, which together consume almost half the heroin consumed in the world. Afghan heroin follows three main distribution routes: the Balkan route (transiting Iran, Pakistan, Turkey, and southeastern Europe) to Western Europe, the northern route (through the Central Asian republics) to Russia, and the southern route via Pakistan to the world (e.g., Iran, China, eastern and southern Africa, etc.) by air and sea (UNODC 2010). In the last decade there has been a boom in heroin consumption in East Africa, an explosion in cocaine use in West Africa and Latin America, and an increase in

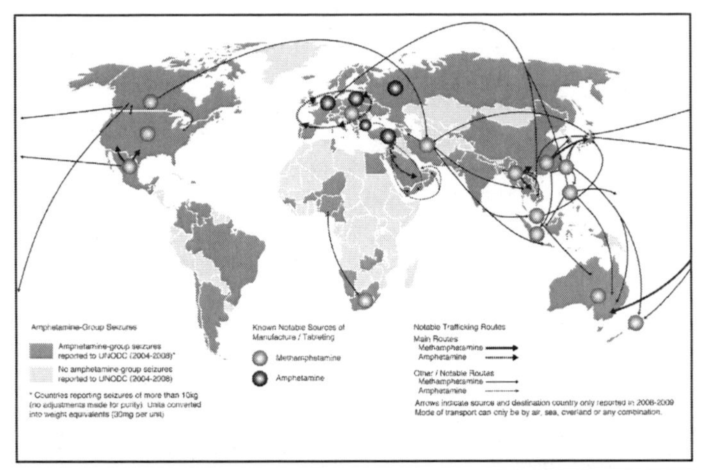

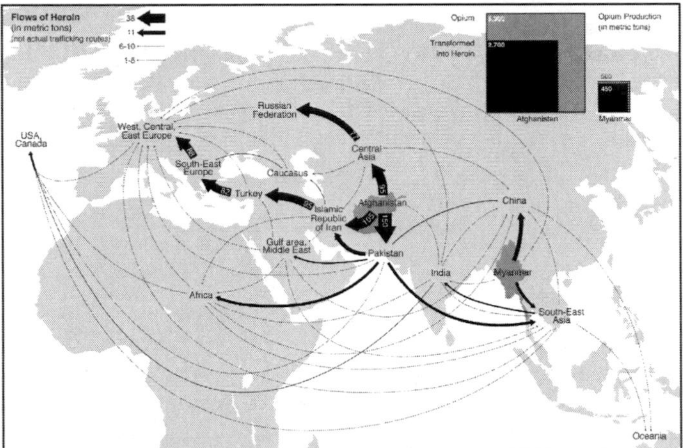

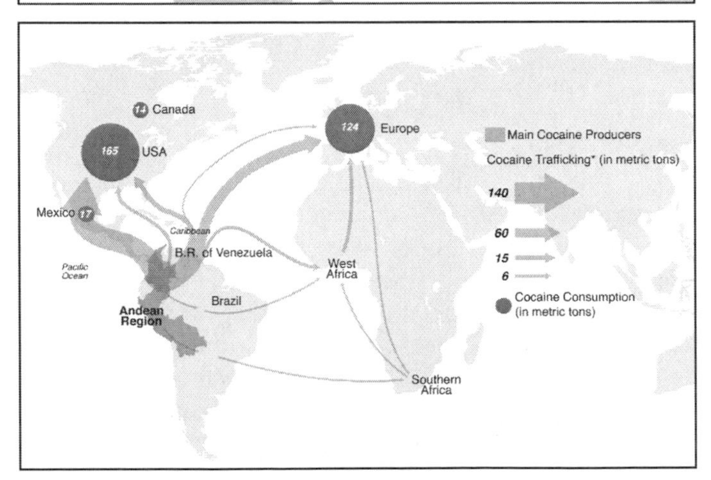

Figure 2.1. The transnational drug market. Top: notable locations of manufacture and main trafficking routes of amphetamine-group substances (2008–2009); center: global heroin flows of Asian origin (in metric tons); bottom: main global cocaine flows in 2008 (in metric tons).

Source: UNODC (2010, 45, 70, 100)

synthetic drug abuse in the Middle East and Southeast Asia. The "problem drug users" have been moving from the developed to the developing world, contributing to higher crime rates, violence, corruption, and the spread of HIV/AIDS. Drug trafficking threatens political instability on transit areas in Latin America and West Africa.

Vignette one illustrates both the rise of a new actor in world politics, the drug trafficker, and the challenge this actor and the drug business poses to the world's established national boundaries, using a review of the film *Traffic* (2000). In this case the territorial border still matters, but as a challenge to be overcome surreptitiously and behind which affluent users are clustered, rather than as a barrier that can only be breached militarily as in classical geopolitical terms.

VIGNETTE ONE: DRUG TRAFFICKERS HAVE NO RESPECT FOR NATIONAL BORDERS

The hip and polemical film Traffic *(2000) artfully asks, exactly where is the front line in the drug war? Where has the line between the protagonists and the antagonists, good and bad, ally and enemy, been drawn? Is it the militarized line that nominally separates the United States from Mexico, as is popularly imagined, or is the battlefront circumscribed on a more personal scale, in the family, at a bathroom door?*

The film's director, Steven Soderbergh, an avowed critic of U.S. drug policy, in this movie, a spin-off of a British production of the same name, critically explores America's two-and-half-decade-long war on drugs. He clearly makes the statement that current efforts and strategies are ineffective and futile, for not only is the demand for illegal drugs at an all-time high but the supply chains seem infinite and ubiquitous, difficult to identify and impossible to eradicate.

William Weld, the former governor of Massachusetts, in a cameo appearance as an ambiguous political figure in the film, offers some wisdom about the nation's drug problem: "You'll never solve this on the supply side." Weld's comment at a Washington, DC, cocktail party to the new "drug czar," played by Michael Douglas, is suggestive of one of the movie's central themes: that despite spending almost 18 billion dollars at the time of filming, the U.S. war on drugs is a failure, and that it has failed mostly because the supply can never be effectively cut off, no matter how much money and blood go into the effort. It's an "unbeatable market force" according to the fictional character Eddie Ruiz, the midlevel dealer who agrees to become an informer after being caught by federal agents. "It's a lot easier to get than alcohol" is the mantra of the young characters, including Caroline Wakefield, the sixteen-year-old addict whose father ironically happens to be the newly appointed drug czar.

Yet none of these stark realities is fully admitted as the film's drug czar pursues the national policy of border fortification, enhanced import and customs security, and militarized interdiction. Simultaneously, Soderbergh reveals the futility of the putative war down to the trench level, as the idealist San Diego detective played by Don Cheadle tirelessly pursues the suppliers, including a wealthy "businessman" living in a posh San Diego suburb, even after witnessing the assassination of his partner as well as a key informant.

On the other side of the divide, Mexico is depicted in broad sepia tones as that dark place where corruption flows unimpeded and where an honest cop is about as rare as moderation in a crack house. Yet one Tijuana cop, played by Benicio del Toro, somehow manages to remain incorruptible and even heroic as he doggedly battles the drug cartels and the complicit government. Despite seeing his friend and partner murdered, and his own life standing on the razor's edge, his selfless and pragmatic approach is ultimately vindicated when, at the end of the film, a little league baseball field is illuminated through his "sincere" efforts. The symbolism of that lighted field amid the dark chaos of the drug war is a powerful message for those back in the real world who advocate a less combative approach.

In numerous interviews, Soderbergh insists the movie wasn't made to change drug policy, but he also thinks that the time is ripe for change, and that the movie will help the process along. "I feel absolutely that it's in the air right now," Soderbergh told a reporter as Traffic *neared the $60 million box-office mark after just six weeks.*

He's right about the timing. Strange things are happening in the drug-policy realm that seemed politically impossible just a short time ago. Nine states have passed ballot measures legalizing the use of marijuana for medical purposes. Many states have proposed lightening up on draconian drug-sentencing laws. A national drug czar from the late 1990s, General Barry McCaffrey, urges that we stop calling the antidrug campaign a "war," reinforcing the remark of Michael Douglas's character in the closing scene of the film: "I don't know how you wage war on your own family."

If the seeming futility of the war on drugs calls the meaning of many national boundaries into question, there are other contemporary political situations that appear to validate their continuing significance and appeal to varied groups, even when on closer attention the opposite may be the case. Vignette two examines one of the world's most intractable conflicts, paying particular attention to the difficulty of establishing "statehood" for Palestine without providing contiguous territory over which a new state could have sovereignty. Yet the economic reality of the situation is that a new Palestine, wherever it is located, will be dependent on Israel for a wide range of services and much employment for its population. Perhaps the only way out of the impasse in the long run is for both sides to accept a single state within which power would be shared? As yet, however, we are very far from that eventuality.

VIGNETTE TWO: WHY ISRAEL AND PALESTINE CANNOT COEXIST ON THE SAME TERRITORY

Edward Said, the late noted literary scholar and Middle East commentator, offers a frank account of the demoralizing aspect of the seemingly immutable and intractable Israeli-Palestinian conflict. In his opinion, it is mostly about the inability to understand and embrace each side's narrative or story about why they claim the same territory:

> *We [Palestinians] were dispossessed and uprooted in 1948; they [Israelis] think they won independence and that the means were just. We recall that the land we left and the territories we are trying to liberate from military occupation are all part of our national patrimony; they think it is theirs by Biblical fiat and diasporic affiliation. Today, by any conceivable standards, we are the victims of the violence; they think they are. There is simply no common ground, no common narrative, no possible area for genuine reconciliation. Our claims are mutually exclusive. Even the notion of a common life shared in the same small piece of land is unthinkable. Each of us thinks of separation, perhaps even of isolating and forgetting the other. (Said 2000)*

This sobering commentary suggests that the Oslo Accords of 1993 and the resulting "peace process," predicated on an anachronistic political-geographic fix to intermingled populations claiming the same territory, has done more damage to the peace and reconciliation efforts in general and to the Palestinian people in particular. In Said's estimation, the occupation of Gaza and the West Bank of the Jordan have gone on too long (since 1967), and the peace talks have dragged on with too little to show for them. The Palestinian goal, even if it is now a "separate" Palestinian state, seems no closer, whatever the bravura about taking their declaration of independence to the United Nations in 2011, and the suffering of ordinary people has gone further than can be endured. Consequently, the outside world continues to witness the vicious circle of stone throwing in the streets and the renewed support and revival of the Intifada as well as retrenched Israeli exceptionalism and expansionism under the sign of Zionism. Furthermore, media misrepresentation has made it almost impossible for the American and European publics to understand the detailed geographical basis of the events, in this, the most geographical of contests.

The only hope, according to Said and others, is to keep trying to rely on an idea of coexistence between two peoples in one land. This codependency idea allows for deliberative and discursive forms of democratic participation that are not predetermined by membership in separate, delineated territorial states. It is a way of confronting the problem within political spaces as opposed to across boundaries and borders—the latter having been the modus operandi of the twentieth-century state-centric world. The intractability of the conflict suggests that a new, less state-centered critical perspective is required—one that does not privilege the state and the discourse of solving sovereignty issues by territorially dividing up the world.

Yet, because each side is "territory mad," determined to have exclusive rights to the territory of the other, no real solution is in sight. The fiction of Palestinian sovereignty inscribed in the Oslo Accords is visible to all but the most cynical Israeli and most naive American politicians.

Ever since the election of Ariel Sharon as Prime Minister in February 2001, achieved in large part as a result of the absolute boycott of the election by Israeli Arabs, and with subsequent governments outdoing the others in expanding settlements in hitherto largely Palestinian areas of the occupied West Bank, the political mood within Israel in general and in relation to the "peace process" in particular has reflected despair, overt acrimonious separatist sentiments by Palestinians, and the further entrenchment of and justification for a state permanently under martial law. In other words, popular opinion, on both sides of the conflict, is such that complete partition into separate political territories, no matter how untenable, as will be shown, is the preferred solution. One is prone to challenge this underlying logic by wondering if it is possible or feasible to address and solve one of the world's oldest ethnic conflicts by merely drawing the appropriate boundaries in political space.

So what of this vaunted and much celebrated "peace process," which at its core has tried to territorially divide the region? What has it achieved and for whom? Why, if indeed it was a peace process, have the miserable and desperate condition of the Palestinians and the loss of life become so much worse than before the signing of the Oslo Accords in September 1993? What does it mean to speak of peace if Israeli troops and settlements are still present in large numbers? For instance, according to the authoritative Report on Israeli Settlement in the Occupied Territories (RISOT), the rate of settlement building has doubled since 1993 and more than 305,000 Jews live "illegally" (as of 2010) and against the dictums of the accords in the West Bank, even if Gaza has been abandoned for the cause. Before Oslo supposedly "forbade" this, the figure was much lower (480 in 1948, 112,000 in 1993). Furthermore, the RISOT does not factor in the 192,000-plus Israeli Jews who have taken up residence in Arab East Jerusalem under the rubric of Former Deputy Mayor of Jerusalem Abraham Kehila's racist doctrine of "Judaification" (2,300 in 1948, 153,000 in 1993). Kehila is quoted as saying in 1993, "I want to make the Palestinians open their eyes to reality and understand that the unification of Jerusalem under Israeli sovereignty is irreversible." Has the world been deluded, or has the rhetoric of "peace" through boundary making, exemplified by the Oslo (1993), Cairo (1994), Taba (1995), Wye (1998), Sharm-el-Sheik (1999), and so on agreements, been essentially a gigantic fraud?

To begin with, we provide some elementary facts about the case. In 1948 Israel took over most of what was historical or British Mandated Palestine, destroying and depopulating 531 Arab villages in the process. Two-thirds of the population was removed or forced out—these are the origins of the four million refugees of today. The West Bank and Gaza, however, went to Jordan and Egypt respectively. Both were subsequently lost to Israel in 1967 and remain under its control to this *day, except for a few areas that operate under a highly circumscribed Palestinian autonomy—the topographical specifications of these areas were decided unilaterally by Israel, as the Oslo process specifies. In other words, Israel took 78 percent of Palestine in 1948 and the remaining 22 percent in 1967. It is only that 22 percent that is in question now, and it excludes West Jerusalem, which had been conceded in advance to Israel at Camp David.*

What land, then, in accordance with the hopes of the peace process, has Israel returned? According to Edward Said, "it is impossible to detail in any straightforward way—impossible by design." It is part of Oslo's design that even Israel's concessions were so heavily encumbered with conditions, qualifications, and entitlements that the Palestinians could not feel that they enjoyed any semblance of self-determination. As the influential New York Times *columnist and Middle East expert Thomas Friedman remarked, "Israeli propaganda that the Palestinians mostly rule themselves in the West Bank is fatuous nonsense. . . . Sure, the Palestinians control their own towns, but the Israelis control all the roads connecting these towns and therefore all their movements."*

It is the constructed geographical map of the peace process that most dramatically shows the distortions that have been developed and have been systematically disguised by the calculated discourse of peace and bilateral negotiations. The Oslo strategy was to redivide and subdivide an already divided Palestinian territory into three subzones—A, B, and C—in ways entirely devised and controlled by the Israeli side. "Palestinians themselves," Said notes, "have until recently been mapless." They had no detailed maps of their own at Oslo, nor were there any individuals on the negotiating team familiar with the geography of the Occupied Territories to contest decisions or to provide alternative plans. Subsequently, this lack of geographic knowledge coupled with Israel's claims of biblical manifest destiny created the noncontiguous territorial arrangement. The resultant tensions have likewise justified the Israeli Defense Forces' continued control.

In the lexicon of the three subzones (see figure 2.2), A *refers to the area of full Palestinian self-rule,* B *to partial Palestinian self-rule, and* C *to Israeli security and civil control. Any one of the maps reveals that not only are the various parts of area A separated from each other, but they are surrounded by area B and, more vitally, area C. In area B, Israel has allowed the Palestinian Authority to help police the main village areas, near where settlements are constantly under construction. Despite the nominal sharing of police powers, Israel essentially holds all the security cards in area B. In area C, it has kept all the territory for itself, 60 percent of the West Bank, in order to build more settlements, open up more roads, and establish military staging grounds. Jeff Halper views this cleverly intended setup as a "matrix of control from which the Palestinians would never be free."*

The Gaza component of area A was much larger mainly because, with its arid lands and overpopulated and rebellious masses, it was considered a liability for Israeli settlement, which was happy to jettison all but the best agricultural lands. Israel finally jettisoned Gaza as a whole in 2005 by demolishing all of its settle-

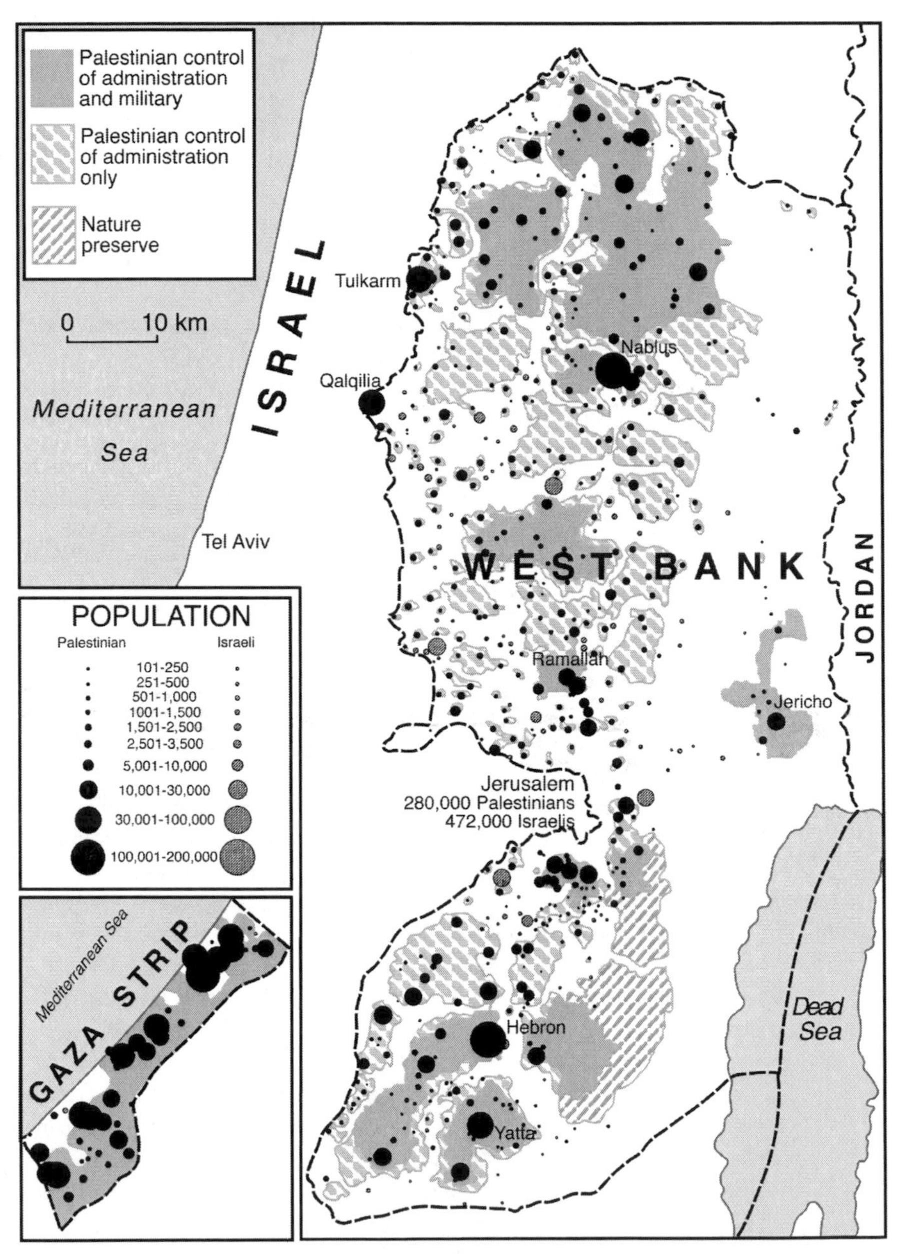

Figure 2.2. Population map of the West Bank and Gaza, showing Palestinian and Israeli settlements and the three zones of the Oslo Accords.

Source: Redrawn from Margalit (2001, 22)

ments. In the West Bank, however, the closures and encirclements mandated by the three subzones have intentionally turned the Palestinian lands into besieged pockmarks on the map.

In October 2000, Amira Hass, a correspondent for the Israeli newspaper Haaretz *in the Palestinian territories, succinctly summarized the situation, using the language of South Africa under apartheid. Effectively, a further eleven years on from when these words were written, nothing has changed other than the Israeli construction of a "fence" and greater incursion of Israeli settlements into what was in 1993 "Palestinian territory":*

> *More than seven years have gone by, and Israel has security and administrative control of 61.2 per cent of the West Bank and about 20 per cent of the Gaza Strip [area C], and security control over another 26.8 per cent of the West Bank [area B]. This control is what has enabled Israel to double the number of settlers in ten years, to enlarge the settlements, to continue its discriminatory policy of cutting back water quotas for three million Palestinians, to prevent Palestinian development in most of the area of the West Bank, and to seal an entire nation into restricted areas, imprisoned in a network of bypass roads meant for Jews only. During these days of strict internal restriction of movement in the West Bank, one can see how carefully each road was planned: so that 200,000 Jews have freedom of movement and about three million Palestinians are locked into their Bantustans until they submit to Israeli demands.*

Furthermore, added to Hass's observations is the fact that the main aquifers for Israel's water supply are under the occupied West Bank; that the "entire nation" of Palestine excludes the four million refugees who are categorically denied the right of return, even though any Jew anywhere in the world enjoys an absolute right of return at any time; that restriction of movement is as severe in Gaza as it is on the West Bank; and that Hass's figure of 200,000 (now 305,000) Jews in the West Bank enjoying freedom of movement does not include the 150,000 (now 192,000) new Israeli Jewish inhabitants who have been brought in to "Judaize" East Jerusalem.

To exacerbate the conundrum, the slow pace of the unfolding peace process is justified by the U.S. and Israel in terms of safeguarding the latter's security; whereas, one hears nothing about Palestinian security or protection from a colonizing aggressor. "Clearly," Said concludes, "as Zionist discourse has always stipulated, the very existence of Palestinians, no matter how confined or disempowered, constitutes a racial and religious threat to Israel's security."

In conclusion, the understanding of the world political map, as promulgated in Cold War rhetoric and propaganda from 1947 to 1989, was of a set of fixed entities made up of self-contained, territorially distinct, and contiguous states that could be arbitrarily used as "chess pieces," "dominoes," or "billiard balls" in the bipolar game of ideological warfare between the U.S. and its allies on the one side and the Soviet Union and its allies on the other. Whatever ethno-territorial conflicts that emerged during this period—for example, Pakistan/India, Czecho-

slovakia/Hungary, Yugoslavia, Palestine/Israel, and pretty much all of postcolonial Africa—were either suppressed by linking them to the overriding global conflict or addressed using a formula of redrawing the borders—partition—simply changing the lines on the map to fit the demands. It was believed, via the doctrine of self-determination of nations, that territorial contiguity of all ethnic-nationalist claims was not only possible but feasible. All that was needed to fit all the disparate groups in the world into neat, bounded, and distinct pockets was a map and the political wherewithal to make the "reality" conform to the map. Inherited from the older European imperialism, this nonchalance about remapping to achieve territorial contiguity still defines the dominant approach of the Great Powers, not least the United States, to resolving ethno-national conflicts, in the Middle East, the former Yugoslavia, and elsewhere.

Now, at the beginning of the third Christian millennium, it appears that simple deterritorializing and reterritorializing of the world's ethno-national cleavages along the lines of the state-centered sovereignty paradigm no longer works and absolutely does not provide a lasting framework for egalitarian justice, if it ever really did. This is partly a result of the inability to define and categorize, de facto and de jure, the aspiring "nations," as well as the impossibility of circumscribing neat little contiguous circles around these "nations." To assume that every contending group occupies specific, clearly outlined territories was and is fantastically naive and tragic. Using the ongoing Palestinian-Israeli debacle as an example, we can see that efforts to neatly draw boundaries around the two claimants, in accordance with the dominant themes of the Oslo Accords and the resulting "peace process," has proven beyond the imagination of both sides, masking Israel's continued occupation and ongoing colonization of "Palestinian" lands and the economic and political marginalization of "Palestinian" Arabs.

Sources: Said (2000); Halper (2000); Sontag (2001); Malley and Agha (2001); Agnew (1989); Christison (1999); Newman (2002); Yiftachel and Yacobi (2002); Gregory (2004); Shatz (2011).

The Israel-Palestine conflict obviously has its own peculiarities, not the least that Israel is itself a "solution" to the persecution of Jews in Europe and elsewhere and Palestine has no history as a state before the advent of Israel. The territorial claims in this case are also more than simply ethnic or national given that the religious sites in Jerusalem—the Western Wall for Jews, the Al Aqsa mosque for Muslim Arabs—are symbolically central to the impasse between the two sides. The main weakness of the Palestinians is that they do not have a state, and Israel is ill-disposed to give them a real one, even if that made sense, as Edward Said suggests it does not. The present "offer" is for a scattered reservation on mainly barren land.

The fact, then, is that statehood still counts. This is why groups from Quebec to Baluchistan and points in between want their own. It counts both because states gain recognition for national differences that otherwise remain subjugated or unappreciated by others and because the world is still largely organized in terms of states for a wide range of activities, from postal and monetary systems to welfare and military organization. Three vignettes illustrate these contentions. Vignette three uses a particular example from Eastern Europe to make the point about national recognition, nascent and possibly fruitless as the process may be in this case. Vignette four makes the rather different point that statehood exists virtually as well as actually, in the sense that de facto recognition of a state-mandated activity, in this case postage stamps, endows a claimant state with a degree of legitimacy. In other words, statehood is as much about external recognition as it is about internal organization. It is a social as much as a material fact. The example of stamps issued by the Armenian enclave of Nagorno-Karabagh in the Caucasus Mountains of Azerbaijan is used as an illustration of the vitality of the idea of statehood in a global context where it is still highly valued. The strangeness of the two cases suggests, however, that in a world in which statehood can be so easily expected or implicitly recognized (if questionably realized), it is something different from the statehood exercised by Great Powers or successful welfare states. Of course, geographical variability in the relative powers of states both within and outside their boundaries has long been a feature of the world political map. As the number of states proliferates, it is one that should not be forgotten. We are now up to 193 and still counting. Some of these states are hardly much of anything in terms of their effectiveness on the ground. Vignette five addresses one aspect of the conundrum of the world's possibly most failed state: Somalia. This is the explosion of piracy off its coasts as the ability of the central government of Somalia to police its territory has collapsed and as alternative means of existence have been invented by people whose traditional livelihoods (such as fishing) have collapsed along with statehood. Vignette five suggests that statehood can be more fragile than much of the dominant discourse in political geography and other fields usually contends.

VIGNETTE THREE: HAIL RUTHENIA! ETHNIC/REGIONAL REVIVALS AND REVOLTS

"Sub-Carpathian Rusyns, arise from your deep slumber"; and so begins the anthem of the aspiring Ruthenian nation. The What? Who? Where? The place is Ruthenia, located in a wedgelike position surrounded by Ukraine in the east, Slovakia and Hungary to the west, Poland to the north, and Romania on the southern flank. The Ruthenians, or Rusyns, are a part of the family of east Slavic peoples—akin to Russians, Belarusians, and Ukrainians—who all of their modern "history" have been controlled by neighbors. They are mainly farmers or woodcutters in the heavily forested Carpathian foothills, and as Timothy Garton Ash

Figure 2.3. Ruthenia in its regional setting.
Source: Garton Ash (1999, 54)

chides, "everything about their origins, culture, language and politics is disputed." Farcical or not, they and their intellectual representatives are clamoring for attention and more autonomy within the current Ukrainian state, itself a recent re-creation. And with individuals spread throughout the region, as well as the United States, prompting the self-proclaimed moniker of "the Kurds of Central Europe," the Ruthenians' demands for their own state takes us to the heart of one of the most important problems of contemporary international politics. In the decades following the Cold War, in an environment of rediscovered freedom and political liberalism, suppressed and sometimes only half-formed nationalities have reemerged and formulated political aspirations all over Europe.

The Ruthenian story is, in every respect, a quintessentially Eastern European one. Yet in Western Europe, as well, there are nationalities, in varying degrees of formation, that strive for anything from autonomy to statehood—Scottish, Welsh, Catalan, and Basque. Further, such claims are not simply endemic to Europe. UNPO, the Unrepresented Nations and People's Organization, maintains a website that lists up to fifty such entities throughout the world, including Abkhazia, Aboriginals of Australia, Baluchistan, the Kurds, Tibet, and Kosova. It is a salient issue in dictatorships as well as liberal democracies, with varying degrees of violence involved. One of the big questions that Ruthenia prompts for Europe is whether the ethnically checkered successor states of the former Soviet Union, beyond the case of the Caucasus, might yet still go the bloody way of the former Yugoslavia.

For most of their modern history, Ruthenians lived in the Austro-Hungarian Empire. When the empire was broken up after World War I, they found themselves scattered between Poland, Hungary, Romania, Yugoslavia, and the Soviet Union, but the greatest concentration was in the new state of Czechoslovakia. Czechoslovakia, the most democratic and liberal of those successor states, gave them considerable linguistic and political autonomy, in a province it called Sub-Carpathian Rus. During World War II it, as an appendage of Slovakia, was the pawn of Hitler and the Hungarians, and it finally ended up in Stalin's Soviet Union as an oblast, or region, of Ukraine. When the Soviet Union collapsed, it became part of the new Ukrainian state.

What makes the Ruthenian issue particularly significant in contemporary discourse is that they, according to Samuel Huntington's (1993) thesis about a coming clash of civilizations, straddle two of the great dividing lines in Europe—one religious, the other geopolitical. The religious split is between Western (Catholic or Protestant) and Eastern (Orthodox) Christianity. The Ruthenians worship in either the Orthodox Church or the Uniate (or Greek Catholic) Church, which uses the Eastern rite but acknowledges the authority of the Western pope. In geopolitical terms, the Ruthenians are on the edge of the new eastern frontier of NATO, with sizable minorities in member countries Hungary, Poland, and Slovakia. Furthermore, as the European Union has expanded, Ruthenia now sits directly on that entity's eastern border. It does not take a great stretch of the imagination to envision an independent Ruthenian state, recently separated from a directionless and unstable Ukraine, aspiring for entry into both NATO and the EU.

The Ruthenian situation is still far from that of the Kurds or Kosovars. For now, their demands are for basic minority rights, like education in their own language. They demand that Ruthenian be an option in future Ukrainian censuses and that Ukrainian state forestry companies stop the mechanized stripping of the trees from their beloved hills—a symbol of their national heritage. They further hope to prevent the Trans-Carpathian oblast from being incorporated into a new, enlarged province governed from Lvov. They also continue to look for more cooperation across the frontiers, in what is already the Carpathian Euroregion. Improvements in a Slovakia trying to appease the Eurocrats of Brussels on issues of minority rights will assuredly increase the grievances in neighboring Ukraine.

The twentieth-century history of the Ruthenians, and their experience in the geopolitics of the region, can be perfectly surmised by a popular Eastern European joke. The joke tells the story of an old man who says he was born in Austria-Hungary, went to school in Czechoslovakia, married in Hungary, worked most of his life in the Soviet Union, and now lives in Ukraine. "Traveled a lot?" asks an interviewer. "No, I never moved from Mukachevo." What isn't a joke, however, is the Ruthenian aspiration for statehood within the context of a very economically and politically unstable region dealing with the ideological residues of the Cold War. The issue of statehood is sure to develop as the Rusyns continue to "arise from their slumber."

Source: Garton Ash (1999).

VIGNETTE FOUR: STAMP COLLECTING AND THE RECOGNITION OF STATES

Invoking William Shakespeare's rhetorical question, "What's in a name?" one may similarly ask, "What's in a stamp?" Only the most naive would believe that stamps are solely the necessary tax that is paid to process one's mail. Stamps and stamp production are unabashedly infused with political-geographic meaning and, quite often, are actively produced symbols employed directly, and covertly, as seemingly banal tools in every state's "nationalizing" project. Stamp designing is rooted in specific places, infused with political and psychological intent, and almost always meant to convey territorially specific, quasi-mythical stories and celebratory messages of achievement. The designs may range from the state flag to an indigenous species of plant, a long-dead war hero, a national sporting event, or a popular monument. Whatever the iconography of the particular stamp, the intent is hardly arbitrary.

Correspondingly, the political imagining of stamp production is played out in stamp collecting. Delving specifically into the insular world of the philatelist, one finds a "hobby" directly and indirectly engaged with the more salient and controversial issues of contemporary political geography—state building, ethnic political autonomy, nationalism, international relations, sovereignty. Stamp collectors, in general, believe that stamps should be "legitimate," in the sense that they are potentially valuable as a collectable, if they are recognized internationally in practice, even if they are not recognized expressly, as by treaty or international agreement. Fundamentally, this is the same principle of international law that applies to the recognition of nation-states. A nation becomes a state when the international community begins treating it as such. State sovereignty is not simply proclaimed; it is a creative endeavor valorized by reciprocal relations.

Because the philatelic community is always on guard for not just the rare, valuable stamp but also the fake—the illegitimate, unrecognized, and nonsanctioned pretender—it plays a role in international political discourse. When examined through a critical political-geographic lens, stamp collecting is revealed to be more than just a benign, leisurely pursuit. It is a highly politicized act that not only lends legitimacy to the modern, territorial-bounded-state paradigm but also provides a discourse and a forum for the ubiquitous struggle of national secessionist and autonomy movements. In the dialectic of stamp production and collection, the synthesis is a credibility vehicle for autonomous movements and nation-state claims.

Having one's stamp officially recognized by the Universal Postal Union (UPU) or the venerable philatelist magazine Scott's Catalogue *is akin to recognition by the United Nations. Consequently, creating a stamp to "process mail," with all the requisite semiotics, is one more device employed by aspiring nations seeking sovereignty recognition in the international community. The Nagorno-Karabagh region, high in the Caucasus Mountains of Central Asia, typifies a common scenario played out between the world of international politics and the world of stamp production and collecting.*

Nagorno Karabagh is a fertile but mountainous area of 4,400 square kilometers in the southern Caucasus situated inside what is today internationally recognized as Azerbaijan. The name itself, a Russian-Turkish-Persian compound, is proof of the region's complex and variegated history. It means "Mountainous Black Garden." The Karabagh Armenians call the region Artsakh, or "Strong Forest."

During the Soviet era, Karabagh was a semiautonomous region or oblast, *despite being located entirely within the republic of Azerbaijan, because of its largely ethnically Armenian population. When the USSR began to collapse in 1988, Karabagh declared its independence from Azerbaijan and, subsequently, clamored for diplomatic and institutional recognition from Moscow. Despite the general trend and positive climate for republic-level partitioning at the time, Moscow thwarted Karabagh's claims for autonomy and instead relegated the issue to that of an Azerbaijani internal affair. Yet in 1991 the citizens of Karabagh voted for independence from the USSR, thus exacerbating civil tension. The conflict gradually escalated. The Azerbaijanis besieged Stepanakert, Karabagh's capital,*

Figure 2.4. Nagorno-Karabagh.
Source: Redrawn from O'Lear (2001, 306)

in 1991–1992 and occupied most of the region. Then the Armenians, supportive of Karabaghi irredentism, counterattacked and by 1993–1994 had seized almost the entire region. Some six hundred thousand Azeri (Azerbaijani) refugees were displaced. A Russian-brokered cease-fire was imposed in May 1994, by which time as many as twenty-five thousand people had died and countless more displaced. Under the cease-fire agreement, Karabagh is self-governing and democratically elects a president and an assembly of representatives. While a number of cease-fire violations have occurred since then (the last one being in the second half of 2010), a number of attempts at resolving the dispute have continued and are still ongoing as of June 2011. Meanwhile, Karabagh has its own armed forces and a diplomatic "representative" in neighboring Armenia. Furthermore, it issues visas and prints postage stamps for its mail system (see figure 2.5).

However, according to the two highly respected philatelist publications, Scott's Catalogue *and* Linn's Stamp News, *this is not enough to make the stamps of Karabagh legitimate, because "Karabagh is not a country." Conventional wisdom in stamp-collecting circles maintains that in order to have a legitimate, internationally recognized stamp, you have to be an internationally recognized "country." Hence, aficionados and collectors have rebuked Karabagh's stamps as nothing more than a "Cinderella." A label that masquerades as a "real" stamp from a "real" country is called a Cinderella—purportedly because it's not what it appears to be, but is rather an ephemeral fantasy.* Linn's Stamp News *publishes a list of phony Cinderella countries as a warning to collectors. There are about four hundred countries on their list, from Alexandria and Atlantis to Zenovia and Zulia, including various former Soviet republics whose stamp printing is viewed as nothing more than the pursuit of profit—a scam to acquire hard currency from gullible stamp collectors. However, Matthew Karanian, writing in the* American

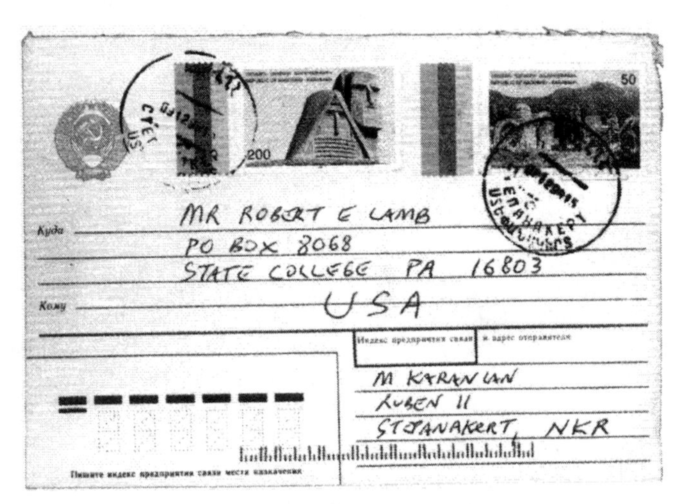

Figure 2.5. Nagorno-Karabagh stamp.
Source: Karanian (2000, 265)

Philatelist, *argues that because the stamps of Karabagh were used to prepay mailing fees, and because they have been accepted by the postal administrations of other countries, the stamps are, in fact, quite legitimate. Further, because they are issued in denominations that are needed to pay actual postal rates, and in reasonable quantities, they do not exploit collectors and, subsequently, should be accorded the respect of other non-Cinderella-produced stamps.*

To process its mail, Karabagh's only alternative would be to rely upon a foreign postal administration and to use the stamps of a foreign country, such as Armenia or Azerbaijan. However, as stated, for twenty years Karabagh has been fighting a secessionist war with Azerbaijan, which, not surprisingly, severely limits postal reliability. Because of the religious connection of Armenian Christianity, Karabagh does get assistance from neighboring Armenia. The mail is trucked to Armenia via a dizzying mountainous road, and from there it is now processed as if it were Armenian mail—with the addition of Armenian stamps.

Although it exacerbates tensions with Azerbaijan, the arrangement is perfectly legal and not unique. A similar situation has existed between Turkey and its beneficiary, the Turkish Republic of Northern Cyprus—also a region undergoing a secessionist feud. Although the "international community" does not recognize Northern Cyprus as a sovereign state, its mail does get delivered. Palestinians also are using their own postage stamps in the occupied territories of Gaza and the West Bank. Mail bearing Palestinian Authority stamps that is internationally destined is sent to and handled by Egypt and Jordan. The international community recognizes the legality of Palestinian stamps, even though Palestine is still not recognized internationally as a sovereign nation-state. Scott's Catalogue *even added these Palestinian stamps to their own catalogue of "legitimate" collectibles in 2000—an event that the* New York Times *deemed significant enough to report.*

Scott's Catalogue *refuses to list Karabagh. The international community does not officially recognize Karabagh as a nation-state. It is not a member of the United Nations, and it has foreign relations only with Armenia—which also does not "officially" recognize Karabagh. But the separatist region does have the international community's de facto recognition of its postal administration. This is shown every time its mail gets delivered outside its borders. And, as Karanian observes, "this recognition certainly should advance the cause of getting Karabagh stamps listed in the stamp catalogues" (266). It is not a great leap of faith to envision how such recognition contributes to Karabagh's state-building efforts. It is a de facto state today, talks about reintegrating it into Azerbaijan having broken down in 2011, so that as time goes by it will become easier to achieve outside recognition, the basis for de jure status. Most recently what has happened is that images of and the name* Karabagh *now appear on Armenian stamps! This is in fact what the Karabagh irredentists wanted all along: recognition of their territory as an exclave of Armenia, not the stamps in themselves. But the stamps have helped keep the cause alive by putting the name and concept of Karabagh in wider circulation.*

Source: Karanian (2000).

VIGNETTE FIVE: PIRACY AND THE FAILED STATE OF SOMALIA

One of the strangest features of the recent past has been the revival of piracy as a major global phenomenon. We usually think of it as characteristic of the early modern period when predominantly English and Dutch sailors captured and ransomed Spanish ships (see Thomson 1994), or of China's shores during the nineteenth century as the imperial government lost its grip over the country's maritime fringe. Though not unique to the Horn of Africa, it has been off the coast of Somalia in the Arabian Sea and around the entrance to the Red Sea that the revival has been most intense. Piracy is symbolic of either states hiring privateers to do their dirty work (as with Elizabethan England) or of the breakdown of central order (as in late-imperial China). Somali piracy very much fits that latter designation.

According to the International Maritime Bureau Piracy Reporting Center, in the first six months of 2011 there were 243 incidents of piracy and armed robbery at sea in the world, 154 of which originated from Somalia. In 2005 just a handful of ships were attacked, and the year after just a single incident was reported. Since then, Somali piracy has risen to 50 attacks in 2007, 111 in 2008, 204 in 2009, and 219 in 2010. Increasing also is the amount paid by the shipping companies in ransom. In 2007 the average ransom was of hundreds of thousands of dollars, while in 2011 it was over $5 million, with the result that piracy is now considered the second-largest source of income in Somalia (after the remittances from the Somali diaspora) with over $200 million a year (Middleton 2011). In late 2008 the EU started the Atalanta operation, soon followed by the Combined Task Force 151 (CTF-151), established by over twenty countries (including Australia, Canada, Greece, the Netherlands, Saudi Arabia, Singapore, South Korea, Pakistan, Spain, the United Kingdom, the United States, Turkey, and others) to patrol an area of approximately two million square miles including the Gulf of Aden and off the east coast of Somalia. These efforts are coordinated with the independent navies of China, Russia, Malaysia, Japan, and India. In response to such efforts, the pirates have expanded their range to the north, east, and south of the Somali coast.

Piracy initially developed at two main locations: Eyl in Puntland (the second breakaway unrecognized state in Somalia since 1998, the other being Somaliland in the north) and Harardhere in the central region. If, in the latter, piracy resulted directly from the complete absence of the state, in Eyl it developed as a consequence of the local administration's decision to use the local fishing fleet to fight illegal uncontrolled exploitation of Somali marine resources by many foreign fleets (from Africa, Asia, and Europe), which, following the 1991 collapse of the state, was thriving. When in 2007–2008 the Puntland administration stopped paying for this "coast guard," many of its members reverted to piracy, which consequently boomed.

If the early attacks were concentrated about fifty nautical miles from the coast, the use of "mother ships," often previously seized vessels, supplied with water and fuel from which to launch attacks has allowed pirates to operate at a greater distance from their shores, sometimes even beyond one thousand miles: to the north—in the Red Sea, the waters off Oman, the Arabian Sea, the Indian Ocean, and up the west coast of India and the Maldives—and to the south—off the coasts of Kenya, Tanzania, Madagascar, and Mozambique. Though the Indian Navy captured about 120 pirates in early 2011, this did not reduce the number of attacks, and a legal framework to address the phenomenon is still lacking. Where should captured pirates be taken and charged after they are in custody? While the initial response to the problem was to increase the number of warships patrolling the area, the rapid adaptability of pirates shows that this is not enough, especially considering that apparently busy fishermen can suddenly turn into pirates when a target approaches.

According to the World Development Report 2011, *the direct economic costs of maritime piracy are estimated at between $6 and $11 billion, including redemptions, insurance, and rerouting, while global efforts to contain and deter the phenomenon have been estimated at between $1.7 and $4.5 billion in 2010. Piracy has therefore become very much a business, which benefits many people, including security companies. It is believed that many of the "instigators" of the pirates are abroad, and only 30 percent of the proceeds from redemptions in fact finally arrives in Somalia. This is still a huge amount of money, though, considering that the average Somali income per capita is around one dollar a day.*

Two elements are at work in this case of massively increased piracy: the collapse of the Somali state, which needs placing in a historical context, and the availability of tempting targets for capture and ransom offshore for men whose traditional occupations have disappeared. The case, however, is representative of a much larger phenomenon that has only been noticed in recent years despite having been around for some time and still perhaps increasing: the increased number of "ungoverned spaces," in which alternatives to the authority of states have arisen to make up for the corruption and ineffectiveness of states around the world (Clunen and Trinkunas 2010). These range from areas with persistent political insurgencies, such as the southern Philippines and Colombia, to regions dominated by criminal gangs, such as Calabria and suburban Naples in Italy, parts of Brazilian cities, and large swathes of Mexico, and places with major sectarian and clan tensions where the writ of the central government does not hold over large parts of the nominal national territory, such as Lebanon, Iraq, Pakistan, and Afghanistan.

Since the opening of the Suez Canal (1869), the coastline of the Horn of Africa acquired increased importance to the European powers because of its geographical position. At the Berlin Conference of 1884–1885, the British, who were guaranteed a good part of the coast overlooking the Gulf of Aden, favored Italian colonization along the coast facing the Indian Ocean to prevent France from expanding its empire in East Africa beyond a tiny territory (current Djibouti). They also gave way to Ethiopia, an independent African empire for millennia (of Coptic Christian religion), to obtain the Ogaden region, inhabited by ethnic Somali, as a reward for their help during the battle for Aden. The Somali

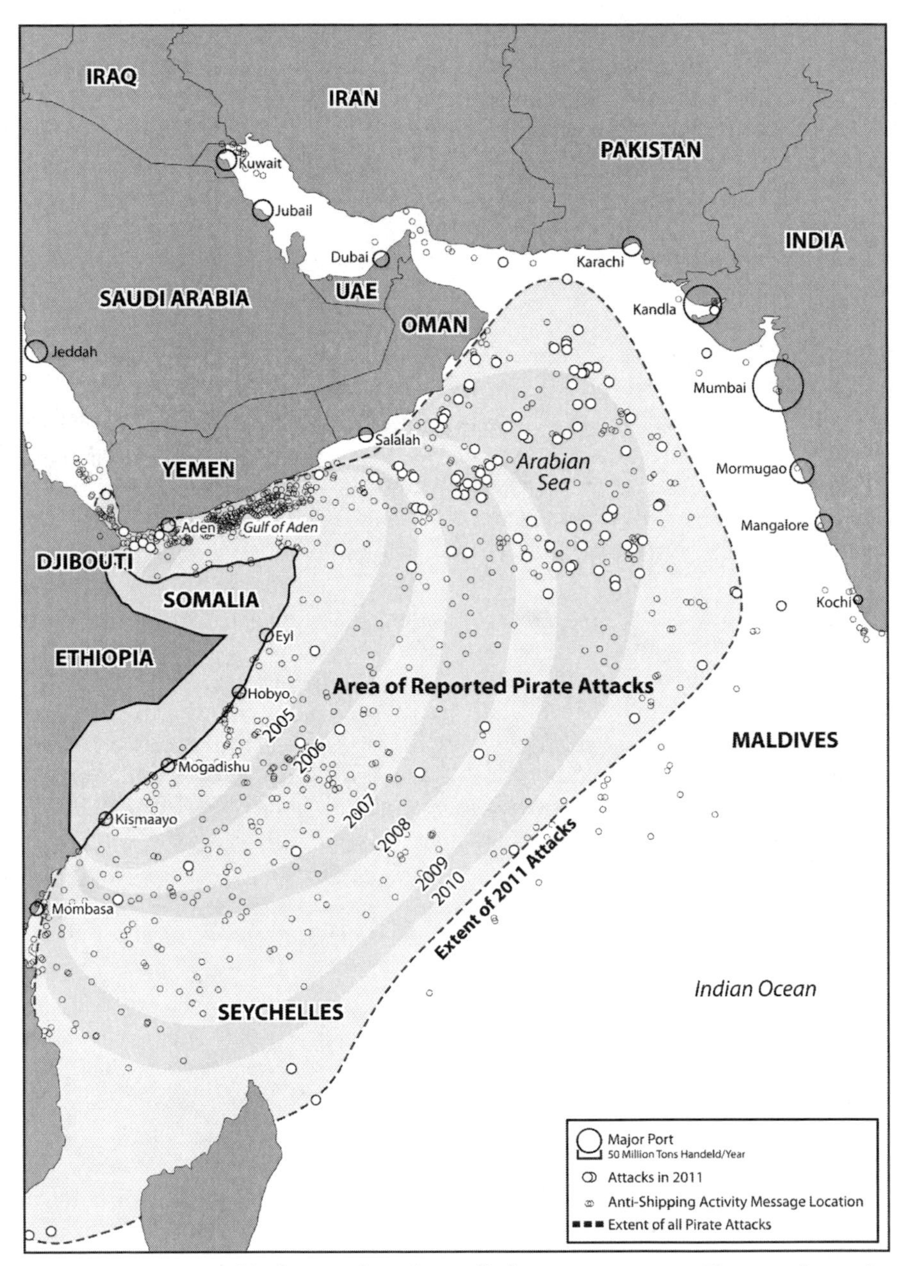

Figure 2.6. Geographical expansion of Somali piracy (2005–2011). The map shows the increasing extent of all pirates attacks and includes data up to March 2011.

Source: Redrawn from Geopolicity (2011)

people were therefore divided among different sovereignties for a long time, with the only exception during the brief existence of Italian East Africa, when Mussolini occupied Ethiopia (1936–1941).

The Republic of Somalia became formally independent in 1960, bringing together in one State the former colonies of Italian Somalia and British Somaliland. Its flag—a blue field with a white star—symbolized the five Somali ethnic groups: British, Italian, and French Somali, plus the Ethiopian Ogadeni and those Somalis in Kenya's Federal District of the northeast, expressing the hope of a future peaceful reunion. In the context of the Cold War, after 1962 the Soviet Union became interested in the new Somali state (Ethiopia was an ally of the United States) and began providing military aid. Within the young state, the 1960s revolved around the debate on how to solve the problem of dependence on foreign aid. In 1969, Siad Barre, general in chief of the army, killed the president and took power in a coup. The effects of economic dependence from abroad, however, continued, despite Barre's policy of "scientific socialism" and attempts at modernization. Meanwhile, in the midseventies, against the backdrop of a devastating drought, Ethiopia changed its regime, now siding with the USSR. Barre tried to take advantage of it by attempting to conquer Ogaden but failed, as the Soviet Union sided against the dictator.

While Somalia gradually turned to the American orbit, the Ogaden defeat marked the beginning of the end of Barre's regime. The dictator's decision to militarize the Somali clans backfired as they began to fragment internally and its rival, Ethiopia, armed the internal opposition to Barre through the Somali National Movement (SNM). In light of its own internal problems, as early as 1988 the Soviet Union withdrew from the Horn. One consequence was that Ethiopia had to withdraw its backing to the SNM, which in turn launched an offensive against the dictator. With the temporary union of two other opposition organizations, Barre—already downsized to the scale of "Mayor of Mogadishu"—was forced to flee the capital in January 1991, in a context of general distraction as all international media were focusing on the First Gulf War in Kuwait and the dissolution of the Soviet Union (Elden, 2009). The collapse of Barre's regime, however, was to prove to be the collapse of the Somali state. The three organizations split, and the SNM, already controlling most of the former British Somaliland, declared the independence of the Republic of Somaliland, though without international recognition. The collapse of Somali territorial unity produced more divisions among "military" factions, often following tribal lines, in a context already strained by previous socioeconomic disputes. While an interclan alliance became impossible, a new drought disaster struck Somalia, and two subclans went to war for the conquest of Mogadishu, symbol and crossroads of national resources, without one being able to prevail over the other.

The humanitarian catastrophe that ensued forced the U.S. to intervene in 1992 with UNOSOM, a multinational force to oversee a temporary truce between factions and further to promote the possible resettlement of a legitimate government

(Operation Restore Hope). UNOSOM, however, turned to political action to eliminate one of the faction leaders, and heavy fighting exploded in the capital. In the 1993 battle of Mogadishu, Somali gunmen hit three Black Hawk helicopters of the U.S. Delta Force, shooting down two. After two days of bloody fighting, five hundred civilians and eighteen U.S. rangers had died. The impact of this event (later portrayed in a movie directed by Ridley Scott), coupled with an inability to reach an agreement, led the UN to declare the mission as having failed and to remove its forces in 1995, leaving Somalia to its own fate.

Even before the demise of the Barre regime in 1995, Somalia had long been in an almost permanent humanitarian emergency. In a region already long stricken by droughts and floods (which climate change has made more frequent in recent decades) causing famines, epidemics, and epizootic diseases, the problems inherited from the colonial and postcolonial times have been immensely worsened by two subsequent decades of civil war since the collapse of the Barre regime. The initial generalized armed conflict that followed between 1991 and 1992 brought an exodus of about three million people, almost half of the Somali people at the time.

According to the World Health Organization, of an estimated population of around seven or eight million inhabitants, about 3.2 million, or approximately 43 percent of the total, as of 2011 require assistance and humanitarian aid for survival. Many are refugees who have fled Mogadishu and south-central Somalia in search of safety. There were nearly ten thousand victims of the civil war that took place in the aftermath of the Ethiopian invasion in December 2006. Fighting in Mogadishu alone in 2007 produced almost four hundred thousand refugees. Their number increased by 77 percent in 2008 alone, and the UN High Commissioner for Refugees (UNHCR) estimates that as of 2011 within Somalia there were about half a million refugees, while almost as many have fled to Kenya, Yemen, Ethiopia, and Djibouti because of war and famine.

As the Somali population continues to suffer the consequences of chronic civil war as the product of Somalia's colonial and postcolonial history, any further attempt at reconstructing the state will hardly succeed, unless more power is given to a civil society that survived for decades in the face of an almost continuous sequence of environmental emergencies, civil war, and humanitarian disasters. In particular, concrete alternatives have to be given to the young people (about half of the Somali population is under twenty-five), who have never seen any peace or security, not to mention evidence of a working state. Among the many responsibilities of the "external actors," there is not just the "policy of identities" that too often legitimized one group at the expense of others. There is also the stigmatizing of the very identity of the Somalis—today as pirates, yesterday as terrorists, anytime as migrants—and explanations of the enduring crisis—clannishness or nationalism— that overlook the fact that these are effects rather than causes (Verhoeven 2009). Religious fanaticism, interclan rivalries, and nationalistic violence are different aspects of the same extreme self-defense reaction by a community feeling threatened.

Originally fishermen attempting to protect the national territorial waters from the abuses of foreign fleets after the collapse of the Barre regime but later joined by a motley crew of armed unemployed men, the pirates have found a path to opportunity by engaging directly with the global economy (Bahadur 2011). For the time being, Somali piracy is not going to stop, especially considering that its costs still seem "affordable" to the shipping companies.

It thus seems unlikely that the problem will be solved by military repression on the high seas. This may only limit the problem. But no lasting solution will be found without tackling the political, economic, and social causes on the land. Helping the local economy and the local fishing industry and rebuilding the social order in Puntland may alleviate it, but the solution will not come until the political problem of Somalia is settled: a problem that is a complex entanglement of historical multiscalar interactions between actors that today interweaves local, national, international, and global scales.

Sources: Verhoeven (2009); Geopolicity (2011); Middleton (2011); Bahadur (2011); Elden (2009); Clunen and Trinkunas (2010).

Political geography, therefore, has increasingly expanded beyond the concern about the territories and borders of states. The world has made us do it. Too many phenomena relating to the political organization (and disorganization) of space no longer conform to the spatial forms of old. The contemporary subfield is now about the geographical distribution of power in general and how this relates to other geographies such as those of statehood, ethnicity, class, gender, and sexual orientation, and how these geographies produce the politics of identities and interests. Even as large parts of the world at large have become more and more integrated, political inequalities between the places occupied by different social groups within countries and cities have become more rather than less marked. The "new" world economy privileges those regions and localities best positioned in relation to other "nodes of advantage," while other places and their residents are left behind. Within global cities such as Los Angeles, London, and Paris, as much as in marginalized regions such as parts of southern Italy or the old textile towns of northern England, are areas of serious economic and social disadvantage in which many people are trapped by a lack of empowerment in relation to local and national political institutions as much as by a lack of employment and employability in the marketplace. If in some parts of the world, such as Karachi in Pakistan, people seem increasingly to think of themselves and act politically as members of ethnic groups (see, e.g., Green 2011), elsewhere ethnicity intersects with social class to produce both residential segregation and condition the politics it engenders. One such place is the subject of the sixth and final vignette. This is the area often known as South Central Los Angeles. Through showing the difficulties of life in this area, the vignette points to the potential role of political activism even within desperately poor places isolated from the main power centers of American society.

This story has wider relevance to the general issue of how power and resistance are implicated in people's lives within the boundaries of states (see, e.g., Herbert 2006, 2011). In this case the relevant boundaries are those of a neighborhood within a major city. But they could equally be the boundaries of regions and localities anywhere that affect the life circumstances and prospects of those living within them. Boundaries matter whenever they demarcate differences in power and opportunity between groups of people in relation to markets and governmental institutions. In the Los Angeles Metropolitan Area, political power is dispersed across a large number of municipalities, special districts, and public agencies. Within the city of Los Angeles—a weirdly shaped entity that sprawls from the San Fernando Valley in the north to the Los Angeles Harbor in the south, the dominant politics is that of ethnicity. The Westside and the Valley are still largely white, the Eastside is now largely Latino, and South Central, largely African American since the 1940s, is increasingly Latino. In the 2001 citywide mayoral election, a white candidate with a mainly white and African American coalition (James Hahn) defeated a Latino candidate with a Latino and white coalition (Antonio Villaraigosa). Villaraigosa gained his revenge at the next election when he too put together a more successful cross-ethnic electoral coalition. But electoral politics hides the fact that there are dramatic differences in class and ethnicity across LA's neighborhoods. LA is a highly segregated city with the quality of public services and private prospects largely determined by where one lives. Where one lives, therefore, is not merely incidental to your life. As numerous empirical studies show, life chances are not simply distributed at the national level (although living in the United States does give just about everyone a head start over Somalia) but more fundamentally at a local scale. Future health, income, and educational, occupational, and political prospects all take root in and around where you live (see, e.g., Newburger, Birch, and Wachter 2011). Neighborhood of residence sets the parameters for life chances and the potential fruitfulness of political action. It is where racial, ethnic, and class differences come together that determines how much power one can exercise in one's own and in other people's lives. This is the political geography of everyday life for ordinary people.

VIGNETTE SIX: WHAT DON'T KILL YOU, MAKE YOU STRONGER: NEIGHBORHOOD, "DESTINY," AND POLITICAL AGENCY

South Central is the part of Los Angeles known elsewhere in the city as its most dangerous and deadly neighborhood. In the year 2000, seventy-six people were killed in the part of South Central covered by the Southeast Division of the Los Angeles Police Department (LAPD), including twenty-three victims under twenty-one years of age. If in 2000 the lion's share of the homicides relative to population in Los Angeles took place in this area (200 out of 544 in the city as a whole), by 2009 things had improved. In 2009 there were forty-one homicide victims in the area compared to 330 in the whole city. Most of the victims in 2000 were young black men. By 2009 the victims had diversified, reflecting both the increased heterogeneity of the population in the area and the diminution in the number of murders that were "hits" ordered by criminal gang bosses. Overwhelmingly, however, the deaths were not random but still the result of gang initiation killings and disputes between gangs and drug dealers over territory. Of course, in both years sometimes innocent people just happened to be in the wrong place at the wrong time. Between 2000 and 2009 the district also lost considerable population. In a neighborhood in which regular employment in factories or other workplaces has become increasingly rare, the drug business fulfills a need for employment and serves the demand for illegal drugs from more affluent neighborhoods. Yet it also

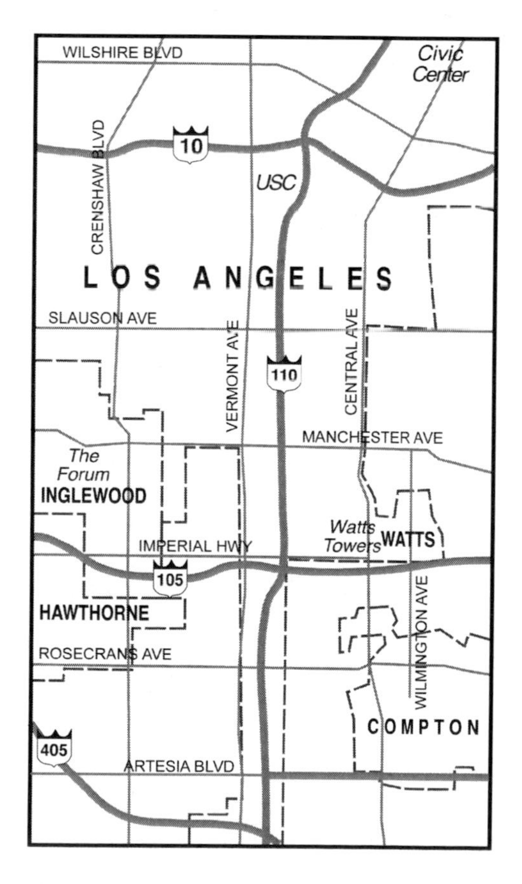

Figure 2.7. The streets and local government boundaries of South Central Los Angeles.
Source: Stewart (2001, 11)

has devastating consequences in terms of lives cut short and high rates of incarceration in California's burgeoning prison sector.

Typically, descriptions of neighborhoods such as South Central stop there. Areas like this are the byword for social pathology. But at the same time that seventy-six young men and boys were killed in 2000,

> *nearly 200 students were graduating from Locke High School, kids were splashing in inflatable plastic pools, God's music sweetened the air above churches along Broadway and beats dropped loud and heavy from the windows of passing cars. People went to work, ran their businesses, raised their children, planned their future. And an old man took his horse for a walk down Avalon Boulevard. "It's all love around here," says 21-year-old William Henagan Jr., an aspiring videographer. And danger. "What don't kill you make you stronger," he says of his experiences. "That's the way I look at it." (Stewart 2001, 10)*

There is, then, a paradox at work. As Jocelyn Stewart reports from the neighborhood, life there produces a "wild mix of images, the best and the worst of people, a river in the desert." But what she sees above all is people still chasing "progress," which means trying to better themselves and their families both individually and collectively. The older inhabitants settled in the area because of discrimination in jobs and housing that kept them out of other districts. Though restrictive covenants, documents that barred selling houses to members of designated groups (Jews, blacks, etc.), were banned in 1948 in California, South Central was by then largely African American and remained so thereafter, isolated and ghettoized within the larger city and metropolis. In the 1960s, riots and the closing of factories, such as those of General Motors and Goodyear in nearby suburbs, undermined the local economy. Between 1970 and 1985, seventy thousand blue-collar jobs left the area. This is when drugs came in; local gangs captured the business and served as distributors to the rest of the city and its suburbs. The negatives of the drug business and the war on drugs concentrated in neighborhoods like South Central. Crack cocaine created a new local market in the 1980s. Gangs grew larger and more violent as the profits increased. More and more young men went off to prison, leaving their absence a palpable feature of the area. But a sense of community persists for some, particularly for those who can connect to older traditions of service and public involvement. Some plant gardens, others advise boys without fathers. The massive immigration from Latin America has also revitalized the neighborhood, if at a lower population density. Though, as Stewart notes, "In this now-mixed community, Latinos and African Americans sometimes settle into cold silence, sometimes erupt in open hostility. But there is also as much common ground" (2001, 16). As a result, there has been a dramatic increase in political action, as people "gather to fight problems, not each other." The local Community Coalition has spent the past twenty years battling local problems, from watching the spread of liquor stores to trying to force school improvements, lobbying to have crack addicts treated rather than imprisoned, and changing the way in which the neighborhood is policed from a militarized to a communal model.

Still, as everyone in South Central knows, "A bullet can find you before you find yourself" (Stewart 2001, 17). The most fateful question one young man can ask another is, "Where you from, homie?" If the question comes from someone in a gang, it is about whose side you are on and what will you kill or die for. But it is also profoundly about geography. It is about living in neighborhoods in which the wider society no longer offers much except policing and demand for the drugs that move through them. The final word goes to Jocelyn Stewart (2001, 35):

> *Imagine living in a community best known for its dead and wounded. If you come from here, you know there is much more: those who plant gardens, fix loose bicycle seats and look out for neighborhood children, those who pay respect to the dead by working hard to keep this community alive; body and soul. If someone asks you to talk about the place you come from, they are the ones you think of: the ones altering the landscape, the ones holding up the sky.*

Source: Stewart (2001).

3

The Historic Canon

Political geography had a history before the term itself came into use in the 1890s. For example, the seventeenth-century Englishman William Petty's idea of "political arithmetic" and his "Political Anatomy of Ireland" can be seen as historical precursors of late nineteenth-century political geography. In mid-eighteenth-century France, Anne-Robert-Jacques Turgot used the term *political geography* to refer to the relationships between the facts of geography, seen as all physical and human features of spatial distribution, and the organization of politics. It is also apparent that many of the great figures in the history of political thought, from the ancient Greeks Plato, Aristotle, and Thucydides to the early modern Florentine Machiavelli and later writers such as Hobbes, Locke, Montesquieu, Turgot, Madison, Rousseau, Hegel, and Marx, had ideas about political territoriality and the effects of geographical location and access to resources on conflict and war that can be regarded as basic elements of political geography. They picked up on the practical realities facing political elites and offered their solutions in the context of the historical periods in which they lived. Thucydides' great work, *The Peloponnesian War,* concerns the two decades of war between Athens and Sparta (431–411 BC) and forms the first example of the use of the opposition between sea and land powers that later political geographers such as Halford Mackinder used as a basic organizing principle. The founders of political geography as such, therefore, could draw upon many centuries of relevant thought to inform their research and writing.

What is clear, however, is that their thinking and that of later writers is closely connected to the times in which they wrote. Though they might all like to claim a timelessness to their work, a transcendence of both the present and the situatedness of their knowledge, they were all both limited and stimulated by the general intellectual culture of their historical contexts. This chapter begins with a general overview of the geopolitical context of the period from 1875 to 1945 in which political geography first developed as an academic field. The period was one of intense rivalry between the Great Powers for global power and influence. A second section identifies an important continuity across the period in the naturalized understanding of knowledge that tended to dominate geography in general and political geography in particular. There are interesting and important differences between countries and over the course of the period in the ideas that tended to dominate among political geographers. So the selection of figures and works for close attention in the making of political geography in this period must offer a range of viewpoints. Inevitably, however, this chapter cannot deal with all of the writings of the period. An attempt is made to examine what could be called a "historic canon," those works and their authors whose intellectual and political influence was greatest during the period and since. But we also pay some attention to ideas and figures whose ideas were less influential. Their experience tells us something about the limits of originality when it fails to connect with the ethos or *zeitgeist* of an era and how it is later remembered. But it also suggests that we should not simply consign everyone to the same intellectual box simply because they lived in the same time period.

We designate one group of figures and exemplary texts as founders: Friedrich Ratzel and Halford Mackinder on the one side as initiators, and Paul Vidal de la Blache and Elisée Reclus on the other as critics. This is followed by subsequent sections on Wilsonianism and American political geography, focusing particularly on Isaiah Bowman; space, race, and expansion, with a focus on the German Karl Haushofer and on the creation of an Italian and a Japanese geopolitics; geopolitics versus political geography, with discussion of texts by Bowman, Albert Demangeon, Yves Marie Goblet, and

Jacques Ancel; and, finally, a discussion of the role of political geographers in World War II. A conclusion sums up the chapter and points the way forward to post–World War II political geography.

THE GEOPOLITICAL CONTEXT, 1875–1945

The late nineteenth century was both the zenith of European empire building and the time when new extra-European Great Powers, the United States and Japan, emerged into global prominence. So not only were the European Great Powers, particularly Britain and France, renewing their colonial activities under stimulus from the colonialism of newly unified Germany and Italy, but Europe was no longer the sole center of global imperialism. At the same time, and as noted by the British political geographer Halford Mackinder in 1903, the relatively easy expansion of worldwide empires into "open spaces" that began with Columbus's voyage in 1492 had come to an end. With the exception of the polar regions, the expansion of any one empire now had to be completely at the expense of the others. More Great Powers and shrinking space for their expansion spelled possible doom for the established and the up-and-coming alike, unless strategies could be fashioned to protect and enhance what each had from the threats posed by the others.

Of course, the "open spaces" were almost invariably occupied, but by peoples who had succumbed to European deceit, military prowess, and disease. As a result of this history, such peoples had long been judged the civilizational inferiors of Europeans, but now, increasingly, they were also categorized as (natural) racial inferiors too. Into a political atmosphere of intense interimperial rivalry, therefore, came a new way of explaining the global political hierarchy based on the increasing acceptance of ideas of environmental and racial determinism. European "success" was henceforward to be explained in terms of the climatic or racial characteristics of Europeans relative to those whom they had conquered or dominated. If in the past Providence had shone down or God had offered a helping hand, now natural characteristics associated with different world regions and the peoples who lived in them were to become popular ways of explaining global political arrangements and offering insights into how best to plan for this or that empire's future success or brilliance.

The enmity between the Great Powers was not based solely on sheer willfulness or the will to power of this or that domestic group, such as the military or armaments manufacturers. The period from 1875 to 1945 was one of fundamental instability in the world political economy because of the arrival of new Great Powers wanting to muscle their way into markets controlled by others and the declining capacity of Britain, the main commercial as well as colonial power, to provide its lending services to the world economy without damaging its colonial position. It is worth saying more about this because it provides the historical context in which modern political geography came into being.

The period 1815–1875 was one in which Britain held the balance of power in Europe, enjoyed a significant edge in sea power that allowed it a coercive role in imposing its trade and monetary policies around the world, and sponsored a set of doctrines—comparative advantage, free trade, and the gold standard—that, though appearing universal, benefited influential interests in Britain. This combination of European Concert and British hegemony elsewhere began to collapse after 1870 once other states with powerful economic and military assets began to challenge Britain. Germany was by far the most important of these. Its capabilities could not be translated into an enhanced global political role without upsetting both the Concert and the global flows of trade and capital centered around Britain. Concurrently, the increased industrial production of the United States and the European states undermined British industrial preeminence and led British business and governments into the use of nontariff barriers and colonial trade to restrict global free trade and price competition. The net result was an erosion of the system of trade and finance centered around Britain and the emergence of a set of competitive imperial states dividing the world into zones based on territorial monopoly.

Another outcome was a polarization of the Great Powers into two increasingly antagonistic groups. One, headed by Britain and France (with tacit American support), was oriented toward maintaining the mix of free trade and imperialism from the previous era. The other, headed by Germany, was concerned with expanding its colonial possessions and challenging British financial dominance. This division was apparent by the 1890s and gained its most famous expression in the Anglo-German naval rivalry, a race between Britain and Germany to see who could build the most and the biggest battleships the fastest.

Interimperial economic rivalry was powerfully fueled by the growing nationalism of the period. The extension of railway systems around national capitals, the increased role of governments in economic activities, and the growth of mass elementary education conspired to produce an enhanced sense of nation-states as "communities of fate." State boundaries seemed to define natural units whose geographical limits were the product of differences in national "vitality" and "capability." This vision did not go unchallenged by, for example, the growing socialist and anarchist movements of the period. Class more than nation was their putative central category. Even they, however, often succumbed to the pervasive nationalism by organizing nationally and seeing empire as a positive rather than a negative phenomenon for their industrial-world constituencies. "Socialism in one country" was well under way before the Russian Revolution of 1917 and the subsequent use of the phrase as a defense of empire within the future Soviet Union.

The late nineteenth century saw a huge expansion and enlargement of the world economy through the new imperialism. Regions inside and outside of Europe became specialized in the production of specific raw materials, food products, stimulants (coffee, tea, opium, etc.), and manufactured goods on a scale unparalleled in world history. Regional industrial specialization in Europe, Japan, and the United States was stimulated by regional specialization in raw material production elsewhere. The worldwide economic depression of 1883–1896, due among other things to decreased profitability in manufacturing as new producers flooded the global marketplace, encouraged a spurt of investment in raw material production. By 1900 not only was most of the world formally bound into colonial empires (as in India, Southeast Asia, and Africa) or under commercial domination by one or more of the Great Powers (as in China and Latin America), but more and more of the world's resources were drawn into a geographically specialized world economy. Bananas in Central America, tea in Ceylon (Sri Lanka), and rubber in Malaya are just three examples of intensive regional commodity specialization from this period.

This was also a period of dramatic technological change. From 1880 to 1914 "a series of sweeping changes in technology and culture created distinctive new modes of thinking about and experiencing time and space" (Kern 1983, 23). Such innovations as the telegraph, the telephone, the automobile, the cinema, the radio, and the assembly line compressed distance, truncated time, and threatened social hierarchies. The global spread of railroads and the invention of the airplane were perhaps the most important challenges to conventional thinking and practice about time and space. The sense of a "closing world," therefore, was neither illusory nor purely the product of the renewed efforts at colonialism.

With hindsight, World War I can be seen as an almost inevitable outcome of the competitive relationship between Germany on the one side and the dominant powers, such as Britain, on the other. Perhaps the war could have been avoided if militarist attitudes and the nationalism of the time had been weaker. The possibility that other powers would enter the war on your behalf also encouraged an initial recklessness that all sides came to rue once the industrialized killing of the war became apparent. Japan and Italy looked for whatever advantage they could find between them, finally tilting away from Germany for a spell. The United States, with its gargantuan national economy, remained divided in its approach to the conflict. Initially joining in the search for colonies during the Spanish-American War of 1898–1900 and finally siding with Britain in 1917, the United States after World War I became divided between those, most importantly President Woodrow Wilson, advocating an internationalism based on accepting the need to negotiate rather than fight over national differences and those, dubbed "isolationists," who counseled American withdrawal from an active global role.

The lessons of World War I, however, were neither well taught nor readily learned. World War II can be seen as a repeat of World War I for many of the same reasons. The treaties following World War I, in particular the Treaty of Versailles, resolved few of the tensions that had underlain it. Rather, they added new sources of hostility between the main European powers, particularly in the form of the prevalent German view that Germany was excessively punished by economic reparations and territorial losses for its role in starting the war. Moreover, the new states that emerged in Eastern Europe introduced not only greater potential for bilateral alliances (such as France and Poland, Britain and Czechoslovakia) against Germany but also a set of enemies for Germany in a region with significant German-speaking populations and extensive German economic interests. The so-called cordon sanitaire or buffer zone of small states that emerged between Germany and Russia in Eastern Europe after the collapse of the Russian, Austrian, and Ottoman Empires, rather than resolving anything, served as a temptation on both sides for later terrifying interventions (figure 3.1).

World War I did produce a number of significant changes in world politics. U.S. intervention had proven militarily decisive. Japan was now recognized as a major Asian power, having already embarked on a strategy of empire building that was to eventually lead to conflict with the U.S. and the European powers in 1941. The collapse of the Czarist regime in Russia in 1917 created a new kind of state based on a state-directed economy. The new state saw itself as threatened by and threatening to the capitalist world economy. Neither Japan nor Soviet Russia was readily incorporated into what was left of the British-dominated world trading and financial system in the interwar period. The United States, after sponsoring the collective security system called the League of Nations in the aftermath of the war, withdrew from its active implementation. All told, therefore, the changes emanating from World War I worked toward rather than away from a future conflict.

The climax came with the remilitarization of the German economy under Nazi rule after 1933. The fact that World War II involved the active alliance of Germany, Italy, and Japan, the three Great Powers with the most autarkic economies and elites most dissatisfied with the global status quo, shows the degree to which interimperial rivalry was the governing process behind the advent of the war. Of course, their subsequent defeat by the U.S., the Soviet Union, and Britain also came to represent the defeat of the colonial approach to global management of wealth and power (see chapter 4).

Figure 3.1. **The so-called *cordon sanitaire* created by the "Big Four" in 1919 was a buffer zone of central Eastern European states claimed to represent the principle of national self-determination but aiming at preventing future expansionism by Germany (eastward) or revolutionary Russia (westward).**

Source: Redrawn from Mackinder (1919)

NATURALIZED KNOWLEDGE

The university field of geography as whole was invented in the late nineteenth century in part as an offshoot of the growth of national geographical societies devoted to exploration, collecting information about exotic peoples, and the opening up of foreign lands to commerce or conquest. The other part of its origins lies in detailed mapping and portrayal of the regions and landscapes of national territories to communicate the material basis of national identity in the burgeoning elementary schools of the era. In this respect, of course, geography was one of a panoply of subjects with ancient roots that were reinvented under their old names to service the needs of statehood and empire building: from anthropology's measuring of physical differences between human groups and literature's capture of national literary genius to history's telling of distinctive and noble national histories. New fields such as sociology, economics, and political science acquired their own niches in the national service.

The basis to knowledge was contested in many of these fields, including geography. Voluntarist understandings of nationhood were still championed, for example, but were increasingly eclipsed by "blood and soil" conceptions. There was now no alternative: nature mandated it. Idealism of both the transcendental, inherited from Hegel in particular, and the pragmatic varieties, important among Americans, also had their supporters, who often mixed it into accounts that also had materialist elements, borrowed from either biology or political economy. Notions of "will," "spirit," and "consciousness" on the part of collective nouns such as nations, classes, and races also coexisted with the claim that such entities are in fact mental constructions rather than real phenomena. There were widespread fears among dominant classes of racial and social "degradation," the sense that the rapid social and economic changes of the nineteenth century threatened the established social order, itself increasingly less aristocratic (landed) and more bourgeois (commercial). An increasingly prestigious and dominant thrust in all of the new "disciplines," however, was toward a "naturalization" of knowledge claims. By this I mean a tendency to explain human and social phenomena largely if not entirely in terms of natural processes, either physical or biological. In other words, they wanted to explain using processes assuredly not of mental construction that lay outside the questionable "human" realm in which values, interests, and identities were all subject to divergent interpretations and hence less amenable to "expert opinion." Later commentators have often "cleaned up" the complexities of this period too much. So although naturalization of knowledge was the dominant thrust, it was frequently combined with other elements to defy simple categorization as "idealist" or "materialist."

Naturalization of knowledge claims had two vital intellectual preconditions. One was the separation of the scientific claim from the subject position of the particular writer. Claims were made to universal knowledge that transcended any particular national, class, gender, or ethnic standpoint. So even as a particular "national interest" was addressed, it was framed by a perspective that put it into the realm of nature rather than that of politics or society. This "view from nowhere" was by no means new, but it was very important to the new university fields in supporting their assertion of expertise and relevance to addressing the problems of the age. The second precondition was preference for the use of arguments drawn from the natural sciences to explain social and political phenomena. Thus the principle of natural selection as proposed by Charles Darwin filtered down into popular culture and into fields such as geography largely in terms of the idea of "survival of the fittest." This not only encouraged adoption of organic conceptions of nation-statehood (the state as a type of living organism) but also of ideas about racial competition, degradation, and dominion. Much of what passed for Social Darwinism, however, was inspired by the older evolutionary ideas of Jean Baptiste Lamarck. These were both more open than Darwin's reliance on variation over extended time periods to the direct effects of the physical environment on social processes and, crucially, to the impact of "will" or intervention in creating more successful organisms. This allowed for the packaging of seemingly contradictory elements into a single study, such as races as biological categories arrayed according to their superior "consciousness," for which there was no natural basis whatsoever. Such ideas were widely shared among elites, not least the new academic ones, across all of the Great Powers. In fact, Darwin's approach to natural change as the root of social order could be used to argue, as did the Russian geographer Peter Kropotkin, that cooperation was more important to evolution than competition, at least insofar as humans are concerned, an understanding revived recently (Nowak with Highfield 2011). Be this as it may, Darwin himself can be used as exhibit A with respect to the limits of naturalized knowledge. Not only can contradictory meanings and even empty claims (Fodor and Piattelli-Palmarini 2010) be found in Darwin's writings; more significantly, in terms of the limits of naturalized knowledge, it is the "inherent instability of Darwin's prose, discernable in his 'syntax as much as his semantics,' which has spawned radically different responses" (Finnegan 2010, 259, quoting Beer 2000, xxxviii; more generally on this theme, see Livingstone 2007).

Modern political geography arose in the conjunction between the practical needs of state and empire building, on the one hand, and the extension of environmental and biological ideas to the realm of world politics, on the other. Political geography, therefore, exhibited both of the traits of the naturalization of knowledge. In the first place, as its proponents argued for how physical geography directed state and empire formation, they also set themselves up as problem solvers for their own states and empires. Thus Mackinder provided a global geopolitical model but was concerned primarily with the implications of the model for the future of the British Empire. That the latter may well have inspired the former rather than vice versa never seems to have occurred to him or anyone else at the time or to some since. Secondly, by transposition from evolutionary biology, the European territorial state acquired the status of an organism with its own needs and demands. But this idea had older and quite different roots. German idealist philosophers from the late eighteenth and early nineteenth centuries, such as Fichte and Hegel, had regarded the state as a being or entity with a life of its own. It was but a short step from this to Ratzel's conception of the state as an organism, a step made possible by the spread of ideas from biology into the emerging social sciences in the 1890s. What lent this further plausibility was the sense that states were now involved in a struggle for survival in a geopolitically "closed" world where one state could now benefit only at the expense of others.

The entire cultural atmosphere of the period conspired to favor the naturalized view of politics and society and made the *physical* in geography especially attractive as a source of explanation of state character and behavior. Four features of the period's dominant strands of thinking in political geography are directly connected to the historical context, both political and intellectual: the harmony of state and nation, natural political boundaries, economic nationalism, and the determining character of location or environmental conditions.

The late nineteenth century was a time of severe social dislocation in Europe. At one and the same time there were massive flows of people within Europe and across the globe, an explosion of urbanization, increasing capital mobility undermining local financial circuits and disrupting established patterns of investment, and strong movements of revolt against these shifts and their impacts in the form of workers' rights and women's suffrage organizations. One response to this was to try to re-create at the national scale the local social harmony that had supposedly prevailed in past times. Just as sociologists, such as the Frenchman Émile Durkheim, argued for a national moral order to replace dying religious and local ones, political geographers in Germany, France, and Britain all participated in the attempt to give the contemporary state an organic character by connecting it to the nation. A mythic community, in which every social stratum knew its place and duty to the whole was projected onto the nation, served as the ideal upon which to build a more stable geopolitical order. An organic local past was to be re-created in the national present. In this order there was no room for cultural differences within the boundaries of states. Every "race" was seen as requiring its proper place. Jews, being culturally distinctive and scattered around Europe, were particularly problematic in this context and as they were identified increasingly in racial rather than in religious or ethnic terms. The deep irony of Durkheim, the assimilated French Jew advocating a new national moral order, is worthy of particular note. Indeed, the German political geographer Ratzel is often identified as the founder of the ecological theory of race in which Jews as a religiously distinctive group are seen as the one *race* most out of place:

> In the Near East they were productive (for example, creating monotheism) but in Europe they have no real cultural meaning. The association of place and race is linked in the rationale of the German in Africa or the Jew in Europe. They are presented as mirror images, for while the German in Africa "heals," the Jew in Europe "infects." (Gilman 1992, 183–84)

This line of thinking, with Jews and other widely diffused and mobile peoples designated as contaminants, was to have devastating consequences after the Nazis gained power in Germany in the 1930s and set about eliminating those who did not fit into their territorialized geopolitical imagination based on racial categorizations (and which today are widely viewed as bogus biologically and thus not "natural" at all).

But whether or not Ratzel himself can be accused of being a proto-Nazi is another question entirely. His emphasis on external environmental circumstances sets him apart from the genetic racism inherent in the Nazi obsession with the German *Volk*. Even more to the point, Ratzel favored German participation in empire building outside of Europe, denying that Germany could expand in Europe itself. Of course, this is hardly reassuring for those outside of Europe whose racial inferiority, even if on environmental rather than genetic grounds, condemned them to inferiority relative to their German (or other European) masters.

A second dominant element in the political geography of the time was the idea that each state has or should have "natural boundaries." This implied, of course, that not all political boundaries were the proper ones. The territorial status quo that had characterized the Concert of Europe was openly called into question. But it also implied that all of the members of a putative nation or ethnic group had the right to live within the boundaries of their own state. Finally, it opened up the possibility of using natural features to designate the natural area of the state. Swedish conservatives, such as Kjellén, argued against Norway's independence from Sweden on the ground that the Scandinavian mountains were not a natural boundary. Seas and rivers were preferred. The arbitrariness of the classification points to the political uses to which such natural claims were put. The Nazi concept of *Lebensraum*, borrowed from Ratzel but now expanded beyond its use as a biogeographical analogy to justify the "right" of vital peoples (such as Germans) to expand into areas not exploited efficiently enough by their current residents, was used to justify subjugating *Mitteleuropa* (Central and Eastern Europe) and was based in the notion of natural boundaries.

A third element in the political geography of the epoch was economic nationalism. The national economy existed to provide for the national community. Ideas of laissez faire and free trade were seen as antithetical to social stability and as sources of social decay. This perspective was widely shared by emerging political elites across Europe and Japan and, to a lesser extent, in the United States. Two of the founders of political geography, the Swede Kjellén and the Englishman Mackinder, both subscribed wholeheartedly to this element of the "organic conservatism" of the era. The nation-state (and its empire, where appropriate) was defined as the basic unit for all economic transactions. Individuals and businesses were held to be subordinate to the greater needs of the nation-state. Writers as different in many respects as the English economist Hobson and the English political geographer Mackinder shared organic definitions of national interest as the driving force behind economic growth. To Hobson, however, empire sacrificed the "home" economy, whereas for Mackinder empire was the means of maintaining the economic basis to military power that was essential to national survival. The Britain that had produced the empire now required it to maintain its global dominance. In the context of the time, Mackinder was probably correct. The long-term problem, of course, was that the discourse of benevolent imperialism rested on the systematic subjugation of vast numbers of people for whom the British maritime empire was anything but an exercise in benevolence. Be that as it may, what Mackinder and others agreed on was the organic unity of a national/imperial ecònomy as an entity in itself.

Economic nationalism was by no means new. In the seventeenth and eighteenth centuries, doctrines of "cameralism" and "mercantilism" had dominated economic discourse. They saw monopoly structures, such as government trading companies and heavy government regulation, as essential for economic growth. The economic growth in question, of course, was that of the nation. By the nineteenth century mercantilism had transmogrified into imperialism because of the difficulties of achieving sustained economic growth within static territorial boundaries. If Britain and France practiced "free trade imperialism" (a peculiar combination of restrictive practices within their empires and a high degree of free trade outside it), the upstart powers such as Germany could only grow by initially engaging in policies that protected their industries by sheltering them from international competition. Then they too could grow economically by expanding territorially, that is, by acquiring a colonial empire.

This too was now seen as biological in nature. As frontiers closed in North America and the world's land masses were all brought into the world market, control over territory appeared to be a crucial requirement for economic growth. The idea of a "closed system" was vital to the plausibility of the biological analogy. Only in such a setting did the zero-sum game of each organic entity competing for a larger slice of the same economic pie make sense. This was what brought nature and nation together. With a "fixed" nature, each nation could only grow at the expense of the others. That the pie could be made larger as a result of innovation and improved productivity was a minority view that never achieved much popularity until after World War II. Indeed, the Great Depression of the 1930s reinforced the logic of competitive national-imperial economies. The state corporatism that became ascendant in the 1920s and 1930s in fascist Italy, Nazi Germany, Spain, and Portugal, freed the state from guaranteeing rights for individuals and groups to pursue a "higher cause," that of guaranteeing the interests of the state (and its dominant elites).

The final feature of the naturalized political geography that developed from the 1890s onward was its emphasis on the determining character of location and environmental conditions. The relative success of different states in global competition was put down to absolute advantages of location and to superior environmental conditions. "Marchland" states (states on the edge of land masses) were seen as possessing intrinsic advantages over "inland" states because they had fewer neighboring states and, hence, fewer potential adversaries. "Maritime" or sea-power states were seen as outflanking "continental" or land-power states in control over the oceans, the main means of global movement. Only the coming of the railroad had called this into question according to Mackinder. This was so because of the relative weight of the Eurasian land mass (or "Heartland") in relation to the difficulty of coalescing and policing the "Insular Crescent" (Mackinder) or "rimland" (Spykman) around its edge. Even those with seemingly impeccable antinaturalist credentials were drawn into the *zeitgeist*. A case in point is the German philosopher Heidegger, who wrote in 1935 of the German *Volk* or nation "trapped" in central Europe:

> We are caught in a pincers. Situated in the center, our *Volk* incurs the severest pressure. It is the *Volk* with the most neighbors and hence the most endangered. With all this, it is the most metaphysical of nations. (Heidegger 1959, 39)

The spatial/environmental determinism associated with formal geopolitical models such as that of Mackinder and later writers such as Spykman, however, was never as popular as a less specific (or more ambiguous) environmental determinism. From this point of view, the Great Power potential of states was a function of their industrial prospects, which, in turn, could be traced to their natural resources (particularly energy resources such as coal) and their ability to exploit them. Some went further and claimed that this "ability" was itself "determined" by climate. From the 1890s until the 1930s such views were not exceptional. Indeed, they formed the mainstream of educated opinion in such fields as geology, geography, and biology. Much of the academic geography of the period in Germany and the English-speaking world involved elaborating systems of environmental/geographical accounting, classifying states and regions in terms of inventories of resources, racial characteristics, economic and political organization, and climatic types. These were widely taught in schools and became the conventional wisdom about why some places "developed" and others lagged behind. Natural attributes determined national destiny.

As we shall see, there were voices that offered criticism of elements of the naturalized discourse upon which political geography was founded, and not all voices always spoke in unison. In particular, some French and American political geographers opposed many features of the conventional wisdom, the French because of their orientation toward a much more historical and human-centered view of the making of states (reflecting perhaps the revolutionary tradition of 1789 to which they were heirs) and the Americans because of their similar "exceptionalist" view of their own past and the liberal-democratic ideals they thought it represented in the world. They tended to have more liberal conceptions of the political and were suspicious of the "determinism" they saw in the pronouncements of figures such as Ratzel and Mackinder. But the relatively high degree of consensus about the conditioning role of location and the physical environment and widespread acceptance of racial categories and colonialism make it possible to characterize the period in intellectually rather homogeneous terms, certainly as compared to more recent times. There is not a little irony in that a period that brought about two world wars was also one in which there was a wide sharing of basic ideas among political geographers in the different countries that spent so much effort preparing for and warring with one another.

THE NATURALIZING FOUNDERS AND THEIR CRITICS

Friedrich Ratzel and *Politische Geographie*

Friedrich Ratzel (1844–1904), like many university geographers of his generation in Germany and elsewhere, was trained in the natural sciences. He studied at the universities of Heidelberg, Jena, and Berlin. After a career as a journalist, he occupied the chair in geography at the University of Leipzig from 1886 until his death. He lived through the process of German unification, and this experience sets him apart from writers who preceded him and those who came after. As a graduate student he was exposed to evolutionary biology in the lectures of Ernst Haeckel and later established a close friendship with the biologist Moritz Wagner, who emphasized the impact of isolation and migration in the creation of species. Space and environmental determinism, then, were central themes of the biology to which Ratzel was exposed. Not surprisingly, therefore, his major work of 1881, the two volume *Anthropogeographie*, placed his new human geography in a naturalistic framework drawn from Wagner's work. His magnum opus, *Politische Geographie* of 1897, was similarly oriented (Livingstone 1992, 199).

Ratzel, however, was also a newspaper correspondent who traveled widely, not least in North America, and wrote prolifically. At his death he left behind a bibliography of around 1,250 items, including 24 books, over 540 articles, more than 600 book reviews, and 146 short biographies (Kasperson and Minghi 1969, 6). Ratzel wrote on a wide range of topics and tended to write in an aggressive manner without much use of caveats or qualifiers. Like many who fancy themselves as "writers" more than communicators, he wrote in a style that sacrificed clarity for effect. This has made his work subject to much controversy over the meaning that can be ascribed to terms that he used and the overall nature of his arguments. He was often inconsistent in his statements about a given term, such as *organism*, for example, sometimes giving a strong biological sense to the term and on other occasions implying a weaker analogy.

Two features of Ratzel's political geography seem clear, however. One is that he proposed to reorganize geography in its entirety around the geography of the state. In this sense his founding of political geography is a refounding of geography as a whole. Places, regions, spaces, landscapes, and all the other concepts that were associated with geographical thinking are now subordinated to the state. The state is "the greatest achievement of man on earth" and the "climax of all phenomena connected to the spread of life" (Ratzel 1923, iv, 2). In this light, his work of 1881, often read as a separate enterprise, inexorably leads to that of 1897. The second feature of Ratzel's work is that the "political" in political geography is restricted to the state. Not for Ratzel was the idea that the political pervades society. The "aristocratic-bourgeois" state (Farinelli 2001) had come to dominate society, and Ratzel acknowledged the tight relationship that thereby existed between geographical knowledge, on the one hand, and statehood, on the other. He was openly acknowledging the role that geography must play in undergirding the "new" state in order to be both relevant and legitimate in the eyes of the state holders. Geographical knowledge must always serve some political project; now was the time of the nation-state.

Something of the flavor of Ratzel's political geography is captured by listing the contents by chapter heading of Ratzel's 1897 book: (1) The

State and the Soil, (2) The Historical Movement and Development of States, (3) The Basic Laws of the Spatial Development of States, (4) The Position, (5) The Space, (6) The Boundary, (7) The Relations between Soils and Water, (8) Mountains and Plains. Unlike most previous political theorists, who had regarded states and other polities as legal-political entities, Ratzel, attracted by the biological reductionism that was all the vogue at the time, conceived of the state as strictly analogous to a living organism, whose territory fluctuated over time depending upon its social and demographic vitality.

Ratzel's political geography is usually presented as organized around seven "laws" of the spatial growth of states. The best version of this perspective in English is a translation of Ratzel's "Die Gesetze des raumlichen Wachstums der Staaten" (Ratzel 1896; translation in Kasperson and Minghi 1969, 17–28). This is more readily understandable than the more extended but elliptical discussion in *Politische Geographie*. Reduced to simple statements, the seven "laws" or tendencies, because that seems to be the sense in which Ratzel is writing, are as follows: (1) the size of a state grows with its culture; (2) the growth of states follows other manifestations of the growth of peoples, which must necessarily precede the growth of the state; (3) the growth of the state proceeds by the annexation of smaller members into the aggregate—at the same time the relationship of the population to the land becomes continuously closer; (4) the boundary is the peripheral organ of the state, the bearer of its growth as well as its fortification, and takes part in all of the transformations of the organism of the state; (5) in its growth the state strives toward the envelopment of politically valuable positions; (6) the first stimuli to the spatial growth of states come to them from outside; (7) the general tendency toward territorial annexation and amalgamation is transmitted from state to state and continually increases in intensity.

Whatever the ambiguities in how Ratzel understood the terms *law* and *organism*, it seems undeniable that his vision of the state as an expanding organism was based on the ideal of a *Grossraum* (large space) as indicative of state dynamism and vitality. Thus he asserted, under the first law, that

> just as the area of the state grows with its culture, so too do we find that at lower levels of civilization peoples are organized in small states. In fact, the further we descend in levels of civilization, the smaller become the states. Thus the size of a state also becomes one of the measures of its cultural level. (Ratzel, in Kasperson and Minghi 1969, 19)

By this reasoning Ratzel conjoined naturalistic reasoning about the imperative of bigness with the political goal of German state building and imperial expansion. Of course, this contradicted the conventional philosophy of the nation-state in its stress on ethnic homogeneity and careful territorial delineation. But it conformed to both the social-safety-valve theory of the settlement frontier provided by Frederick Jackson Turner at the same time in the United States and to the imperialist political mood of the 1890s. Following the logic of Malthus on population and resources but combining this with themes current in evolutionary biology, Ratzel used the biological analogy of the state as an organism that, as its population grew, was subject to resource exhaustion and thus had to expand or die. In other words,

> Ratzel believed he had disclosed the natural laws of the territorial growth of states and he happily located the contemporary thrust of the European powers in Africa as the manifestation of their quest for *Lebensraum* [living space]. Imperial history was the spatial story of a struggle for existence. (Livingstone 1992, 200)

Ratzel's naturalism can be overplayed. His work was as much inspired by the imperialism of the age and euphoria over German unification as by the intellectual attraction to ideas from evolutionary biology. Naturalism provided a convenient shell in which to wrap what was in many respects a remarkably idealist enterprise. This was a point first made, to our knowledge, by Jean Gottmann (1952). *Politische Geographie* and other works are filled with language that could have sprung from the pages of Fichte or Hegel, notwithstanding the equally frequent use of biological metaphors. For example, Ratzel tries to connect the spatial growth of the state to the development of what he calls popular spatial consciousness or space conceptions; "organic" has the sense of the individuality of the *Volk* as much as something akin to an organism; and land and people constitute a natural whole once the people (nation) is rescued from its distress (*Not*) (Dijkink 2001).

Long seen as one of the founders of modern political geography, Ratzel's reputation has suffered through guilt by association with the later German geopoliticians and Nazi expansionism. What connected them was an idealist understanding of the state wrapped in the materialist garb of evolutionary biology. What separated them was the kind of biology they emphasized: the role of external environmental circumstances for Ratzel, genetic racial characteristics for the Nazis (Bassin 1987a, 1987b). Ratzel was very much a product of his time and place and should be understood in this context, not that of the Nazis who later expropriated some of his ideas and terminology for their own purposes (Raffestin 1988; Mercier 1995).

Mackinder and His Geopolitical Model

Halford Mackinder (1861–1947) shared Ratzel's interests in presenting geography as a science that explained human phenomena in naturalized terms and in serving contemporary imperialism. In this case, of course, the imperialism was that of Britain, the status quo Great Power, rather than the upstart Germany. Mackinder used the term *political geography* loosely to refer to human geography in general. This usage is deeply revealing of the degree to which Mackinder, like Ratzel, was committed to a "useful" geography, "geography as an aid to statecraft," which while transcending the nature/culture divide also served the political goal of "reason of state": educating the British population about the world they lived in and the political threats its empire faced because of the rise of hostile powers (particularly Germany) and technological changes (such as the coming of the railroad) that challenged the global dominance of a maritime-commercial empire such as the British.

Mackinder was a publicist for academic geography and a national politician as much as a professional university geographer. Educated in biology at Oxford, Mackinder had a number of important credits to his name: first academic appointee in geography at Oxford University, head of Reading College (later University) (1892–1903), the first ascent of Mount Kenya (1899), director of the London School of Economics (1903–1908), Conservative-Unionist MP (1910–1922), and British high commissioner to southern Russia during the early part of the Russian Civil War (1919–1920). As a figure in the history of geography in general and political geography in particular, he is famous mainly for three publications. The first is an address to the Royal Geographical Society (RGS) in London in 1887 titled "On the Scope and Methods of Geography," which makes a strong case for geography as a "bridge" of understanding between natural and human worlds. This sees the whole of what today would be called human geography as political geography, suggesting that it was in service to the state that geography would overcome its potential bifurcation into two parts. In this respect, Ratzel and Mackinder were equally imperialistic about political geography's position within geography as well as in relation to world politics.

The second paper, also delivered as an address to the RGS, this time in 1904, and the paper which establishes him as a "founder" of political geography, is "The Geographical Pivot of History." Revised on a number of subsequent occasions, this paper argues that sea power was declining in relation to land power as a result of the coming of the railroad, and there was now a pivotal area, "in the closed heart-land of Euro-Asia," that was likely to become the seat of world power (see figures 3.2 and 3.3). Only intervention to keep the pivot out of the hands of a continental power could avoid this outcome. Otherwise, the ready accessibility of the rest of Eurasia to the power controlling the pivot would lead to that power's domination over the rest of the world. This argument had obvious implications for the sustainability of the British Empire, dependent as it was on control over the world's sea lanes and maritime access to India and elsewhere around the edges of the great Eurasian land mass.

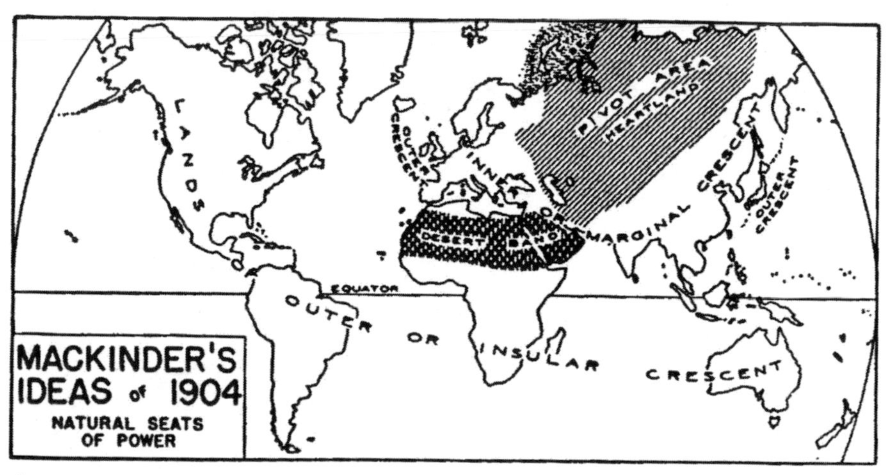

Figure 3.2. Mackinder's "heartland" model.
Source: Mackinder (1904, 424)

The third publication builds a more fully formed political geography in the book *Democratic Ideals and Reality*, published in 1919. Here Central Europe emerges as crucial to the global balance of power. If this developed his already stated pivot idea into a more expanded "heartland" concept, much of the book offered more interesting prognostications about the increasing centralization of political power and the temptation this provided for the rise of "ruthless organizers" committed to manipulating semieducated populations for "non-democratic ends."

Mackinder is a founder of political geography, even though he himself refused to countenance subfields of geography, for several reasons. First, he used physical-geographical conditions, or "realities" as he saw them, to make strong claims about the course and prospects of world politics. Inspired by precursors from the ancient Greek Thucydides to the American naval officer Alfred Mahan, his geopolitical model of sea power versus land power was designed for the ages, even though he used it to make policy prescriptions for the British Empire. This raises the question of the limits of the determinism implicit in the physical "realities" he described. Indeed, rather like Ratzel, Mackinder used naturalized claims to further a more idealist purpose. In Mackinder's case this was "warning" about what would happen to the British Empire if nature was allowed to take its course! Mackinder was a reluctant environmental determinist.

Second, Mackinder was a "man of action." He was not primarily a reflective scholar. Rather, in all his activities he was a reformer, devoted to the "cause" of geography as a subject and to the British Empire as a political-

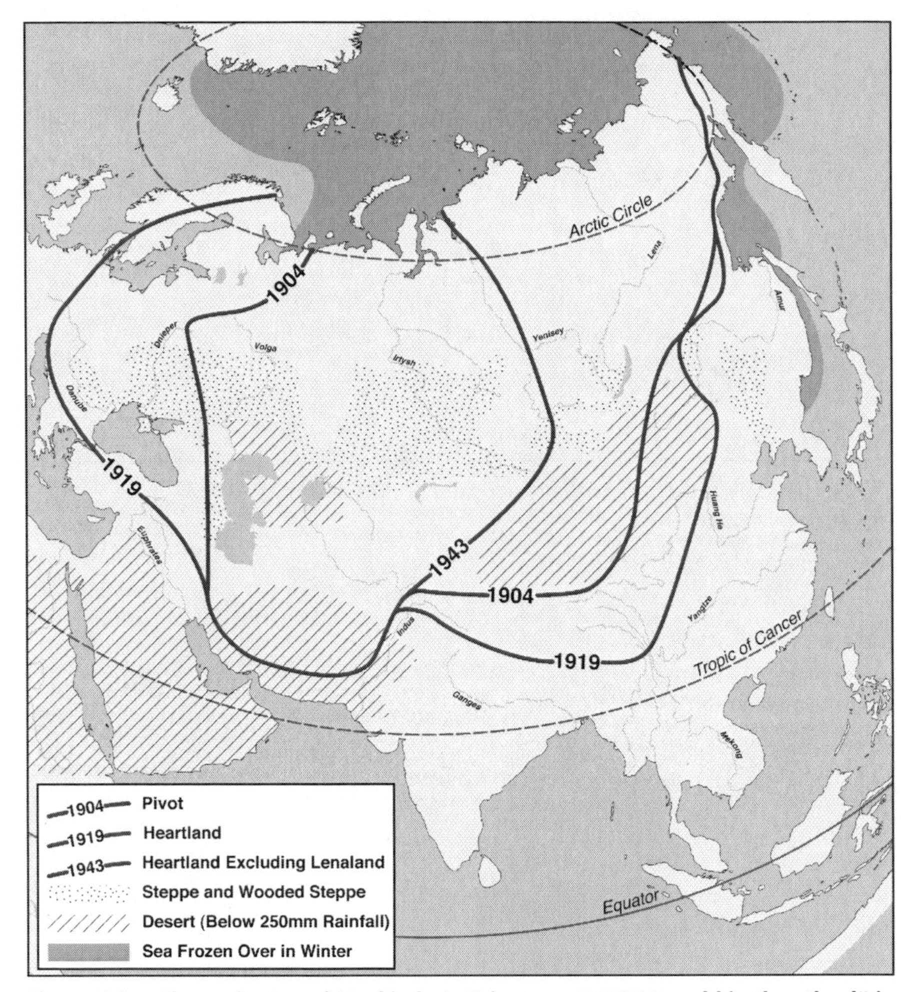

Figure 3.3. The perimeter of Mackinder's "pivot area" (1904) and his "heartland" in the 1919 and 1943 versions of Mackinder's geopolitical model.
Source: Redrawn from G. Parker (1998)

economic project. His writings all have a "manifesto quality" to them, as noted by his biographer Brian Blouet (1987). His major papers were designed for oral delivery to august gatherings. He managed to boil down his messages to simple axioms, such as his famous adage from *Democratic Ideals and Reality* (1919, 150):

> Who rules East Europe commands the Heartland:
> Who rules the Heartland commands the World-Island:
> Who rules the World-Island commands the World.

This combination of political purpose and simple formulas was important in publicizing Mackinder's ideas, which became better known among politicians and nongeographers than those of other contemporary geographers (W. H. Parker 1982; Blouet 1987). Strangely, Mackinder does not receive even a single mention anywhere in a book published in 1994 to arouse interest in the founding connection between geography and empire (Godlewska and Smith 1994). Yet there is hardly a better candidate for making this connection come to life than Halford Mackinder.

Finally, Mackinder's ideas, particularly that of the pivot, have seemed to have a prescient quality to them that later generations have found attractive. They take world politics out of the realm of the historically contingent and into the realm of the geographically predictable. Geography trumps history. Not only was the pivot idea picked up by German writers in the 1920s, if put to very different purpose than Mackinder would have licensed, but after World War II the U.S. policy of "containing" the Soviet Union could be seen as finding its "scientific" basis in Mackinder's heartland thesis (see Gray 1989).

It is a mistake, however, to see Mackinder simply as Ratzel's lost British twin. Even though their views share a common grounding in the biological rhetoric of the time, this is much more muted in Mackinder than in Ratzel. Mackinder was much more oriented to the British Empire as a political-economic concern than as a biological organism (see the earlier discussion of economic nationalism). His "spatial determinism" privileged the distribution of continents and oceans more than climatic or racial characteristics of different areas. Indeed, his emphasis on the pivot could be criticized for ignoring the climatic and resource deficiencies of that region. Thus in his writings Mackinder largely avoids the extreme environmental determinism that Ratzel's American and other disciples (such as Ellen Churchill Semple, Ellsworth Huntington, and Griffith Taylor) made their stock-in-trade. This is not to say that he had anything other than conventional views about race and development. They did not figure prominently, however, in any of his writings. Critics of Mackinder's "imperialism," such as H. J. Fleure, tended to have much more deterministic views of climate and race.

Mackinder also saw himself as a political reformer, using his geopolitical model to proffer political advice to the "Prince." That he was hardly a political radical goes without saying. His notion of "democracy" is hierarchical and paternalistic. But he had an activist view of the state akin to that of social democrats of the time more than to that of the typical conservative. But the statism he espoused was a peculiarly English one. Mackinder was active intellectually at a time when attempts were under way in England to avoid both idealist tendencies to reify the state and individualist (and liberal) tendencies to decry the state altogether (Bentley 1996, 52–53). The new thinking might acknowledge that the state had its territory and that it needed a large one, even an empire. But it also insisted, as did Mackinder, that states are made up of a patchwork of smaller regions or territories with their own characteristics and that the boundaries of the state provided a "rampart" behind which social and political goals could be pursued. In this respect, Mackinder was closer to the thinking of Patrick Geddes, Sidney and Beatrice Webb, H. G. Wells, and other English Fabian reformers than to that of Ratzel and his acolytes. Mackinder's context and that of Ratzel were not the same, therefore, however much we might see similar attempts to integrate the physical and the human by using the organic-evolutionary language of Lamarck and Darwin.

The crucial feature of Mackinder's political geography lost in most discussions of his geopolitical model was the effort to square two aspects of the British Empire that had become increasingly contradictory by the 1890s. After all, his political geography was directly inspired by the cause of empire. On one side was the empire's British aspect: empire as commerce, as maritime power, as democratic or free, and as expansive. On the other was its imperial and transnational aspect: empire as military conquest, as multireligious, as serving Britain, and as based on racial and social hierarchy. Mackinder portrayed the British Empire as a sort of "community of fate" in which all of its residents somehow did or could share in its fruits. But this position was untenable. There was no way to create an imperial identity that integrated the two versions of empire. An enterprise based on social and geographical inequality could not be turned into a homogeneous entity with an all-encompassing purpose. In emphasizing the role of the Other of the pivot or heartland as an external threat, he missed clearly identifying what it was that might keep the empire he so dearly loved together as a viable enterprise. That its disintegration might happen more from within than from without because of the untenability of its imperial identity in the eyes of those colonized by it never seems to have occurred to him.

Vidal's Critical Response

French human geography was founded, it can be said, as a rejection of "political geography" after the style of Ratzel (Robic 1994). Yet in its major early works, such as Paul Vidal de la Blache's *Tableau de la géographie de la France* (1903), the entire focus is on France as a national unit. If anything, Vidal's later work, such as *La France de l'Est* (1917), is even more concerned with the French national question. What accounts for this seeming paradox?

Vidal's approach is important both as a reflection of the time in which he wrote and for the alternative conception of political geography he provided. Contrary to Ratzel and Mackinder, who turned human geography into political geography, Vidal did the opposite. Vidal wrote in the aftermath of France's defeat in the Franco-Prussian War and at a time of intense conflict in France between church and state. He certainly seems to have shared the naturalistic approach to knowledge of his German and English contemporaries, even down to a penchant for Lamarckian ideas of evolution (Archer 1993). He certainly was not the advocate of the "humanistic" perspective that some recent commentators have alleged. But he tended to apply the idea of naturalistic evolution to "regions" or to "civilizations" rather than to states. He also inherited the French tendency, in Dumont's (1983) words, to say, "I am a man by nature and French by accident," whereas the German variant had become, "I am a man because I am a German." This is a profound cultural difference and one that accounts, to this day, for differences between France and Germany in conceptions of citizenship, with the latter having a more ethnic and less "civilizational" (do you speak French?) definition.

The net effect was to realize a conception of political geography antithetical to that of Ratzel. In the first place, Vidal downplayed the importance of national boundaries. Rather, he insisted on the historical contingency and openness of France's boundaries. Indeed, to Vidal, France's peculiarity lay in its fusion of diverse parts rather than as an essential primordial unity. This insight provided the basis to the French tradition of political/electoral geography begun by André Siegfried, one of the founders of modern French political science (of whom more later).

Second, Vidal saw French national identity not in terms of ethnicity or direct environmental constraints but in the fusion of forms of life (*genres de vie*) around national genius loci. This historical process is much more than the history of the extension of political power by the state. It is much more the cultural conversion of "peasants into Frenchmen," to adopt the title of Eugen Weber's (1967) much later book. In this respect, Vidal is the intellectual descendant of the eighteenth-century philosopher Johann Gottfried Herder, who claimed that people bore the "mark" of the country in which they live (Claval 1994). Yet this was not a question of choice. It was how "nature" manifested itself in human life.

Third, Vidal shared the widespread fear among French elites that France was in decline as a Great Power (Heffernan 1994). His culturally expansive view of Frenchness, therefore, could be used to sanction colonial expansion. Of course, the task was thus cast as a *mission civilisatrice* more than a conquest.

Finally, in the *Tableau* Vidal prefers the qualifier "human" to "political" because he believes that the "economic" is becoming more important and the idea of separating out the political from other aspects of life thus makes less sense than it once did. So although Vidal uses "France" as the milieu for his human geography, he is actively denying the total state-centrism that animates the political geography of Ratzel and Mackinder. Notwithstanding the contextual specificities of his work, Vidal offered the beginning of an enlarged conception of the "political"—beyond the geography of the state—even as he opposed the idea of a field of political geography as such.

Reclus: Human Geopolitics?

If Vidal provided a critique of Ratzel's political geography from a French culturalist perspective, there were others who offered more radical perspectives than the dominant views. One such person was Elisée Reclus. A supporter of the revolutionary Paris Commune in 1870, Reclus articulated a much more human-centered geography than that of his contemporaries. In his major work, *L'homme et la terre* (1905–1908), he identified "natural regions," produced by history, language, and lifestyle (*genre de vie*), that are also riven by social conflicts because of inequalities in wealth and power. Class struggle and the increased consciousness of individuals about their capacity to produce change lead to social change. But it is through human agency that inequalities within and between regions can balance out or create equilibrium. It is from "human beings that the creative will to construct and reconstruct the world is born" (Reclus 1905–1908, preface).

Resisting the tendency to reify the state as a separable "thing," Reclus was able to offer a perspective that liberated politics from obsession with the state. The weakness of Reclus's geography lies in failing to see that ignoring the state did not undermine it. Recognizing the rise and importance of nation-states in the contemporary world might have made his perspective more persuasive. As an anarchist, however, his normative rejection of the state led him to downplay its analytical significance. At the same time, he was also compromised by his strange defense of French colonialism in North Africa, even as he railed against imperialism in general (Heffernan 1994). A utopian imperialism, therefore, coexisted with a utopian fixation on a stateless world. This was not a good basis for an alternative political geography, notwithstanding its critique of the global political status quo.

André Siegfried and French Political Science

While geopolitics—despite the academic disputes noted above—achieved a certain academic popularity, the political geography of André Siegfried (1875–1959) enjoyed a less controversial success with the general public, even outside the borders of France, since his numerous books, dedicated to single countries (the U.S., Canada, Great Britain, etc.) or to specific regional aspects (from the Mediterranean to the Panama-Suez maritime corridors), were often translated into many languages. Geographer, political scientist, economist, historian, sociologist, publicist, or simply a social scientist (Gottmann 1987), in his regular columns in the French newspaper

Le Figaro for twenty-five years he contributed a sustained analysis of the Anglo-Saxon world and of the great international problems of his time. Nonetheless, his impact on academic geography has only recently been reassessed (Sanguin 1985, 2010). In fact, Siegfried taught at the Collège de France (1933–1946) and especially at Sciences Po (1910–1959), a school—according to Sanguin—created after France's defeat in the Franco-Prussian War to counter the perceived superiority of German science, and to prepare a new elite for the French government's ranks, following the model of leading U.S. universities' Departments of Government.

This "unconventional French political geographer" (Sanguin 1985) painted his great portraits of the democracies of the time using a comparative method and with an underlying deep social concern:

> His sources are not to be found so much in the schools of Vidal or in German geography, but rather in the tradition stemming from Montesquieu and Tocqueville. And even if in his work on Western France he did follow the regional method of Vidal, there is a much greater accent on the political dispositions of places, which was further developed in his line of research on the psychology of the nations. In this field, with the inspiration coming from Michelet and Tocqueville (on the collective psychology of the American people), he added his own knowledge of economic science, his mastery of statistical tools, in the framework of the great balance of the world's market economy. (Sanguin 2010, 53–54)

Coming as he did from a rich bourgeois family in Normandy (his father, Jules, was a cotton importer with a distinguished political career), the family trade and his early years in the great port city of Le Havre in fact contributed to his openness to the outside world. A great traveler, fascinated by the diversity of cultures and societies, he combined interests in different subjects such as political geography, international relations, electoral sociology, and especially what he called the psychology of the nations. "I've taken the habit of approaching every study as a journey . . . my method is that of the reporter," confronting directly places, people, and institutions, since nothing else could make up for the direct contact and the personal experience of different places. His portraits of different countries were therefore very personal, but he still had a unique ability in capturing the spirit of people and places. Some French geographers reproached him for his distance from theorizing since, as Sanguin observes, "neither in his books nor in his courses did he formulate a doctrine exposing the laws of politics" (Sanguin 2010), while to Claval (1989) "it would not have taken him much effort to pass from such sketches to a more systematic theory of the relationships between power and space." In fact, as Claval (1998) has pointed out, Siegfried belonged to a generation that did not trust either explicit theories or generalizations. In a different light, as Gottmann (1987) wrote, most of his contributions focused on regional aspects, since he saw the variety of the world as too large and complex to allow for quickly elaborated general theories.

Finally, according to Le Lannou (1975), Siegfried's entire work revolves around a geopolitical crisis, that of the change in the relationship between Europe and the rest of the world after the great economic growth of the United States following World War I. Siegfried's seminal works, such as his *Tableau politique de la France de l'Ouest sous la Troisième République* (1913)—triggered by his own experience as a four-times-failed political candidate in Le Havre—and *France, a Study in Nationality* (1930) (published in French as *Tableau des partis en France* [1930]), gained him a reputation for pioneering electoral geography in France, especially among political scientists. Among political geographers, according to Sanguin, his legacy can be found in the work of Jean Gottmann, his assistant for over a decade at Sciences Po, especially in books such as *Megalopolis* (1961) and *The Significance of Territory* (1973); in the electoral sociology of François Goguel (1983), and later, the electoral geography of Yves Lacoste (1986), which reevaluates Siegfried's micro-regional approach; and also, to some extent, in French social geography since the 1960s.

WILSONIANISM AND AMERICAN POLITICAL GEOGRAPHY

Before World War I extreme environmental determinism ruled the roost among most American geographers. Figures such as Ellsworth Huntington and Ellen Churchill Semple represented the establishment in U.S. geography of the naturalized conception of knowledge at its most undiluted. Frederick Jackson Turner's frontier thesis, about how the experience of the European settlement frontier sweeping across North America had made American society distinctive, echoed this type of reasoning. Ideas of racial hierarchy and the need to restrict American immigration for fear of "racial degradation" were widely shared among intellectuals and well-established immigrants, particularly those from Northern and Western Europe. American doctrines and laws governing "race mixing" and eugenics—involving forced sterilization and marriage controls applied to groups deemed "inferior"—struck the German Nazis as useful prototypes for what they had in mind (Kühl 1994).

Though popular among the northeastern elites (and at universities such as Harvard, Yale, and Vassar) because they could be used to justify why "northwest Europeans" ran everything in the country, these ideas did not always sit well with other features of the "American experience." In particular, since the Revolution, with a boost after the Civil War, American political ideology had emphasized two differences between the United

States and the Old World (of Europe) that made environmental (and racial) determinisms both historically and morally problematic (Rosenberg 1982).

The first was an emphasis on free commerce and the ideal of a pacific nation. Here were the main ingredients of American exceptionalism: open frontiers, pacific trade, and no standing armies, only militias. Anchored in an individualistic liberalism, this element of popular thought was antithetical to the statism of contemporary Europe. The second was the idea of the United States as a "social experiment." If not without powerful contradictions, in particular the virulent racism associated with the history of American slavery, this reflected the view of the United States as a place for "fresh starts": people coming from elsewhere and making their lives over or moving around within the country to achieve new beginnings.

Woodrow Wilson, president of the United States from 1913 to 1921, though a white southerner sympathetic to the interpretation of post–Civil War American history portrayed in D. W. Griffith's notorious 1915 pro–Ku Klux Klan film *The Birth of a Nation*, was intellectually and politically committed to the vision of American exceptionalism. People, even leading intellectuals, are not always consistent. Wilson was committed to an American "mission" in the world. He believed that the territorial nature of America was fulfilled, once it had reached the Pacific and defined fairly stable boundaries with Canada and Mexico. America now stood as a "model" for the rest of the world: "the more democracies there were in the world, the wider America's ideological hegemony would spread. A world dominated by liberal capitalism would be the ultimate shield for the American republic" (Perlmutter 1997, 32). The most important reflection of this ideology was to be the League of Nations and a world politics based on collective security enforced by international treaties. But Wilson was no political innocent: "Wilson was one of the first American Machiavellian presidents. He may have seemed naïve, moralistic, and evangelistic, yet he initiated the first American covert actions, and his interventions in Mexico in 1913 and in the Russian Revolution in 1919 demonstrate that this professor of politics from Princeton was no saint" (Perlmutter 1997, 34).

It is in this context that a truly American political geography began. It is associated above all with the figure of Isaiah Bowman (1878–1950). Chief territorial advisor on the American Commission to the Paris Peace Conference in 1919, director of the American Geographical Society, and after 1935 president of Johns Hopkins University, Bowman shifted the focus of political geography from generic, largely speculative arguments about statehood to the empirical structure of state territories. Two influences appear crucial in Bowman's work: the American experience as a model for the world as a whole and the view of World War I as a disaster produced by competitive militarism. In his 1921 book *The New World: Problems in Political Geography*,

Bowman focused on the aftermath of World War I and proposed a common framework of analysis for each of thirty-five countries and world regions (from boundaries and economic conditions to pressing political problems and demography), a gazetteer of politically relevant conditions around the world. This was as much a recapitulation of the old nineteenth-century state geography suited to the new times as it was a reaction against the environmental determinism of Huntington and other American "neo-Ratzelians." In fact, Bowman seems to have been ambiguous about how far to go in abandoning the environmental determinism with which he had been raised at Harvard (Livingstone 1992, 250). But he is important in political geography because for him it was the scientific "neutrality" of naturalism that was most appealing rather than the specific role of this or that biological factor per se. In this regard Bowman is an early American "policy scientist," deploying his knowledge as a problem solver for U.S. "national interests."

This inventory approach developed by Bowman became common in textbooks published from the 1920s until the years after World War II, indicating the lasting influence that Bowman had on the field (see, e.g., Boggs 1940; Pearcy and Fifield 1948). Although written from an American perspective, the focus in Bowman's book on the problems arising from dramatic change in the structural characteristics of states (new boundaries, new ethnic distributions, new communication patterns, etc.) could be seen as practical in nature: offering solutions to real-world problems rather than engaging in theoretical speculation (see extracts below). It thus appealed to the pragmatic imagination of many Americans while fitting into the project of offering a "new world order" in which problems would be solved rationally (i.e., nonmilitarily) by the application of systematic knowledge. During World War II Bowman was a strong proponent of rolling back or limiting European colonialism once the war was over (N. Smith 1994).

Bowman's general orientation was developed further by others such as Derwent Whittlesey (at Harvard), Stephen Jones (at Yale), and Richard Hartshorne (at Wisconsin), with the difference that these figures tended to see "political area studies" more as special cases of regions defined by political processes such as "the impress of central authority" (Whittlesey) and "centripetal and centrifugal forces" (Hartshorne) than in entirely descriptive terms. What unites them is antipathy to the determining as opposed to the conditioning role of physical conditions and a predilection for American-style commercial expansion as opposed to the territorial imperialism of the Europeans.

But an older environmental determinism lived on in some writing, particularly textbooks, in the use of physical-biological analogies applied to world politics. Samuel Van Valkenburg, professor of geography at Clark

EXTRACTS FROM *THE NEW WORLD* (1928 EDITION)
BY ISAIAH BOWMAN

To face the problems of the day, the men who compose the government of the United States need more than native common sense and the desire to deal fairly with others. They need, above all, to give scholarly consideration to the geographical and historical materials that go into the making of that web of fact, relationship, and tradition that we call foreign policy. As we have not a trained and permanent foreign-office staff, our administrative principles are still antiquated. Thus even our loftiest intentions are often defeated. (p. iii)

In the eventual history of the period in which we live, it is reasonable to think that the greatest emphasis will be put not upon the World War or the peace treaties that closed it, searching and complex and revolutionary as their terms proved to be, but rather upon the profound change that took place in the spiritual and mental attitudes of the people that compose this new world. There came into being a critical spirit of inquiry into causes, of challenge to a world inherited from the past, of profound distrust of many existing institutions. The effects of the war were so far-reaching that it was indeed a new world in which men found themselves. (p. 1)

Love of country does not mean hatred for other countries. Patriotism should mean pride in the works of idealism of one's own country. If it has advanced law and order, regional cooperation, international good will; if it has protected the weak, advanced the arts of peace; if its influence has been beneficent—all of these things one can be proud of. But blind patriotism spoiling for a fight is now one of the most dangerous things in the world precisely because the world is now highly organized and war strikes at the very means and spirit of organization and the cooperative process. (pp. 5–6)

University, published a textbook in 1939, dedicated to Isaiah Bowman, containing rather positive opinions about the influence of political geography in the Germany of the time, even though he declares himself "a firm believer in democracy" (Van Valkenburg 1939, viii), giving first billing to the "physical element" in political geography (ahead of economic and human elements), using France as an "example" of political geography with an entire section devoted to "Colonies," and adapting the Harvard physical geographer William Morris Davis's "stage theory" of physical-landscape evolution to "the political pattern of the world" (see figure 3.4). The stage theory seems largely a descriptive device more than a purported explanation, even though Van Valkenburg does use the latter term. After proposing the model, he seems to back off from it (see extract below). One hopes that students of the day noted this lack of enthusiasm.

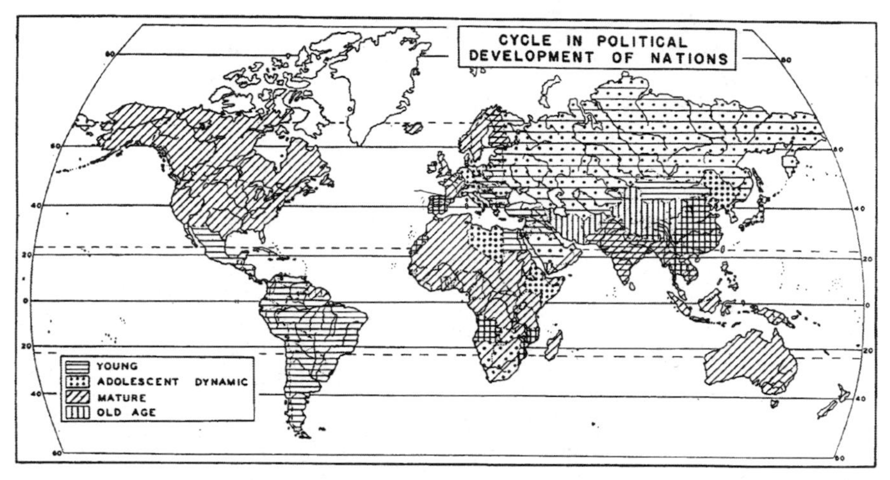

Figure 3.4. Van Valkenburg's "Cycle in Political Development of Nations."
Source: Van Valkenburg (1939, figure 2)

EXTRACT FROM *ELEMENTS OF POLITICAL GEOGRAPHY* (1939) BY SAMUEL VAN VALKENBURG

Is there an explanation for political stability and instability? The world at present offers many examples of countries that want to bring about political changes, while others seem to be satisfied with the existing conditions and defend the status quo. The terms, "have countries" and "have-not countries" are well known, as are also the tendencies of some of the "have-nots" to become "haves." This division of states into two groups is not satisfactory because several "have nots" seem to be perfectly satisfied, while some of the "haves" claim more. The author, perfectly aware of the danger of generalization, nevertheless has the courage to present another explanation for the interrelations between countries and the changes in the political world pattern. This explanation is based on a cycle in the political development of nations, recognizing four stages, namely youth, adolescence, maturity and old age [see figure 3.4].

After completion the cycle may renew itself, possibly with a change in political extension, while the cycle can also be interrupted any time and brought back to a former stage. The time element (the length of a stage) differs greatly from nation to nation and depends on the character of the state; correspondingly no forecast can be made on the time of shift to a next stage. (p. 5)

SPACE, RACE, AND EXPANSION

If the increased physical size of a state was for Ratzel one consequence of greater national "vitality," for Rudolf Kjellén (1864–1922), the Swedish political scientist who first coined the word *geopolitics*, it was the logical outcome of an inevitable competition for power between states. This idea and allied ones, such as the "geopolitical instinct" that linked national populations to their states, attracted the attention of some geographers in post–World War I Germany. Active in nationalist and later Nazi circles, these geographers formed what is often called the German school of geopolitics. Their notion of *Geopolitik* joined Ratzel's organic conception of the state, as refined by Kjellén, to Mackinder's global strategic model. This lethal intellectual brew appealed to Karl Haushofer (1869–1946), the leader of this approach, and his followers because it offered a simple explanation of Germany's plight after World War I and a seeming solution to it. On the one hand, the terms exacted following Germany's defeat ignored the long-term challenge that Germany posed to Britain and the global status quo. On the other hand, the expansion of Germany as a populous state at the expense of less vital neighbors was justified by its need for *Lebensraum*, or living space.

One of the formalized schemes that Haushofer and his colleagues came up with was for combining imperial and colonized peoples within what they called "pan-regions." However fanciful or utopian in terms of the real possibility of ever overcoming the global distribution of power at the time, this type of mapping did express the common assumption that the world was actually made up of racial groupings that could be neatly divided into two "types" of peoples. The one (the colonized) existed to serve the other (the colonizers) (see figure 3.5).

A common tendency toward illustrating the naturalist credentials of what they proposed was the use of maps to show the Darwinian struggle (as they conceived it) between the Great Powers, on the one hand, and the smaller states that constituted both their potential prey and a cordon sanitaire to keep them at bay, on the other. Arrows were used to show the most likely "pressure points" that the Great Powers could use against one another and the smaller states in between. Much political cartography of the time took this *dynamic* form (see figure 3.6).

The political influence of Haushofer and his colleagues, their debt to Ratzel and others, and the role of German political geography in annihilationist discourse directed at Jews and colonized peoples are the subject of considerable debate (e.g., W. D. Smith 1986; Danielsson 2009). Although the Nazi takeover in 1933 and subsequent absolute rule gave prestige to *Geopolitik*, its tenets were not exactly at one with the increasingly strident racism and anti-Semitism of the regime. Indeed, *Geopolitik* looked to alliance with the Soviet Union (realized briefly beginning in 1939 but

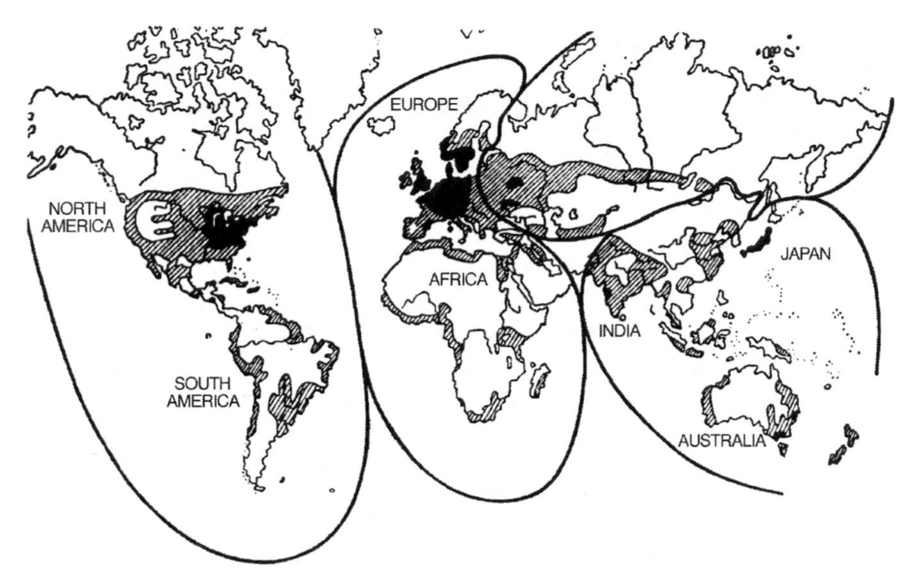

Figure 3.5. Karl Hausohofer's pan-region model.
Source: O'Loughlin and Van der Wusten (1990, 3)

then dashed by the German invasion of the Soviet Union in 1941), saw Central Europe in terms of multiethnic federation, and, at least until the 1930s, paid little or no attention to the "racial question." Even if they laid the groundwork for a strident state-centered logic associated with such concepts as *Lebensraum*, the ideas of Haushofer and his ilk can hardly be seen as lying immediately behind Nazi policy. According to Herwig (1999), Haushofer can be seen as only having an indirect influence on Nazi policies. Indeed, other German geographers, such as Albrecht Penck, can be seen as more important and influential in their emphasis on the need to "unify" the scattered pockets of ethnic German settlement to the east, define "natural" frontiers for Germany more realistic than those of Weimar (postwar) Germany, and turn Germany from the "bourgeois" state it had become into a revitalized *Volkisch Reich*. But even Penck did not offer the lethal mix of anti-Semitism, racial purity, anti-Bolshevism (anti-communism), and *Lebensraum* as a racial concept that Hitler, as the maker and guardian of Nazi thought, came to pursue.

What unified Haushofer on one side and Penck on the other with the Nazis and other German nationalists was resentment about the terms of the Treaty of Versailles and the negative effects they had had on German society, particularly economic reparations and boundary adjustments such as the "Danzig corridor" in Poland and the return of Alsace-Lorraine to France. They used geographical ideas to articulate their nationalist concerns. Penck (1916, 227) put their position most vividly when he wrote, "Knowledge is

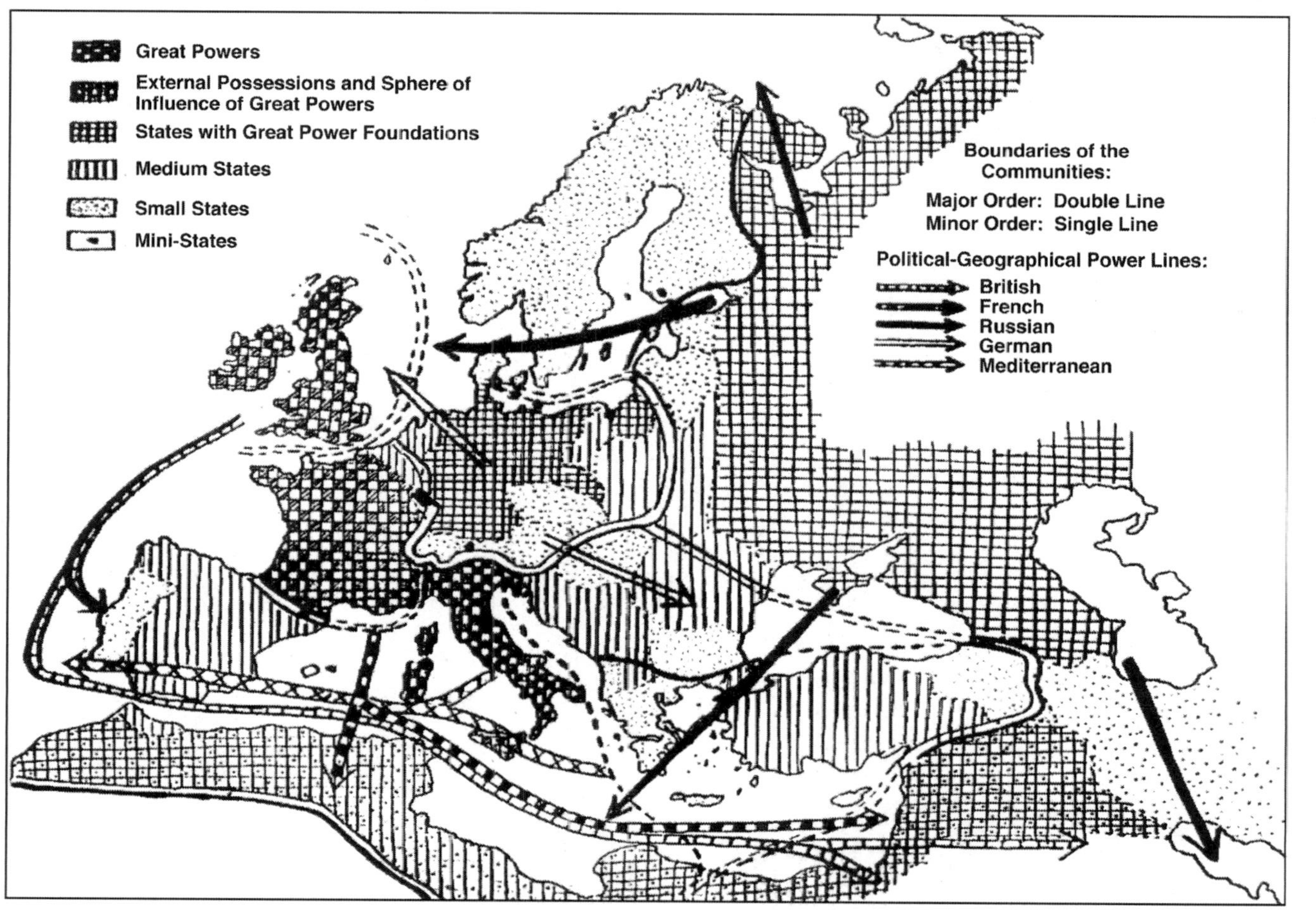

Figure 3.6. Europe as a scene of Darwinian struggle.
Source: Redrawn from Maull (1928)

power, geographical knowledge is world-power." Yet Haushofer also believed, like Ratzel and Mackinder before him, that his geopolitical ideas had a more general validity. In this he was a true heir to the strange alliance of naturalized knowledge and idealist intention apparent in Ratzel's political geography. He did not see himself as simply a propagandist for an expansive Germany. Claude Raffestin (2001) plausibly claims that the Nazi legal theorist Carl Schmitt's friend-enemy political ontology is basic to the geopolitical cartography of Haushofer and his acolytes. The use of geometrical symbols and arrows not only reduces places to points on a plane of "pure politics" (to use a term of Schmitt's) but also universalizes what otherwise would be seen as maps representing Germany's "unique" situation. With respect to the translation of geography into strategy, therefore, the godlike view from nowhere implicit in the maps used by the German geopoliticians represented an attempt to forecast the balance of power between competing states rather than to analyze the geographical contexts of those states in their complexity (Raffestin, 2001) (see, again, figure 3.6).

In other countries, such as Italy and Japan, with similar resentments about their global status, analogous forms of geopolitical thinking also developed in the aftermath of World War I. In Italy, the fascist dictatorship of Mussolini from 1922 on was inspired by resentment at the post–World War I territorial settlement in Europe, the instability of the liberal regime, and the idea of re-creating the ancient Roman Empire anew in twentieth-century Italy (Gambi 1994). As a result, "The new Italy must be prepared confidently to reassert its historic role at the heart of a new Europe" (Heffernan 1998, 139). Interestingly, it took some time for an explicit Italian geopolitics to develop. In fact, only in 1939 did a journal devoted to Italian geopolitics finally appear. Partly this reflected divisions among leading fascists over the relative emphasis to place on the Roman heritage and territorial expansion versus the creation of a more self-sufficient Italian national economy. It also reflected the incoherence of fascism as an ideology based as it was more on what it was opposed to—anti-socialist, anti-democratic, anti-parliamentary—than on what it stood for. But it also represented the fact that Italian fascism aspired to European leadership through the spread of fascism to other European countries. Both German geopolitics and Nazism never had, indeed could not have, this aspiration.

Unlike German *Geopolitik*, Italian geopolitics had a Mediterranean more than a Eurasian and global focus, largely eschewed any kind of racial argument, and tended to emphasize fascism as an ideology that other countries could adopt more than Italian national "superiority." What it shared with the Germans was a style of presentation, particularly the use of suggestive maps casting Italian territorial claims in the most positive light, and an affinity for some of the quasi-mystical elements of the Roman past and its imperial heritage among some leading fascists. One of the main figures in

Italian geopolitics, Ernesto Massi, rehabilitated himself within Italian geography after World War II, becoming president of the *Società Geografica Italiana* (Italian Geographical Society). Though he remained active within the Italian neofascist movement, he adapted to the "new Europe" by becoming an exponent of European unification and Italy's place within it. In fact, the old fascist vision of Europe and the new Europe, shorn of the colonial enterprise, may not be all that far apart, particularly if the European Union fails to adequately democratize. Massi did not have to change that much after all to find a place for himself in postwar Italian political geography. Haushofer would have had a harder time.

Not unlike in Europe, academic geography in Japan began to develop at the end of the nineteenth century, precisely when the country was starting to project its imperial ambitions abroad. Following the Meiji Restoration, teaching geography had come to have a role in building Japan's identity in the popular imagination. But in the context of the heavy borrowing from Western sources that characterized Japan at this time, Western ideas about geopolitics took pride of place. Ratzel had a longstanding interest in Japan (Miyakawa 2001), as did the German geopoliticians of the 1920s. Their ideas quickly spread to Japan, where they took root very strongly, particularly in militarist circles.

Given the enormous loss of life on both sides produced by Japan's 1932 invasion of Manchuria (designed to benefit national industry and deflect attention from internal economic and social troubles), German concepts such as *Lebensraum* provided the government with a handy pretext to oppose domestic concerns and the wide international condemnation the invasion had provoked (Takeuchi 1994). By the late 1930s geopolitics had become popular, with a number of different "schools," such as the Kyoto school, largely oriented toward a distinctive Japanese-focused version of geopolitics that rejected environmental determinism and proposed the creation of a "Greater East Asia Coprosperity Sphere." Another school was made up of academicians who used German geopolitics to justify Japan's imperial expansion. Immediately before the start of the Pacific War, these groups joined forces with others with more critical backgrounds to create the new Japanese Society for Geopolitics, inaugurated in 1941. To Takeuchi, this organization's heterogeneous composition was reflected in the lack of a coherent methodology and ideology. Nevertheless, their arguments seemingly "exerted some influence on the bureaucrats and administrative bodies responsible for wartime policies." The society published a journal that covered topics ranging from spatial aspects of Pacific War strategies and descriptions of occupied areas to "studies conducted in the name of economic efficiency and large-scale regional planning in East Asia's autarkic sphere" (Takeuchi 1994, 200). Takeuchi explains that some of its early critiques were drastically curtailed by the totalitarian nature of the Japanese government. For example, Keishi Ohara was fired from his teaching position in 1937 for, among other things, criticizing the naturalization of geopolitics and its organic view of the state. At the end of World War II, a demilitarized Japan, with its emperor deprived of political power and the country occupied by the United States, purged its most prominent geopoliticians more for their social influence than for any concern about the import of their theories.

GEOPOLITICS VERSUS POLITICAL GEOGRAPHY

Geopolitics, whether that of Mackinder or in its later manifestations as influenced by Ratzel and Kjellén, was never without its critics. To some the "orthodox" interpretations were unsound. The Dutch American political scientist Nicholas Spykman (1944), for example, reversed Mackinder's logic, arguing that control over the "rimland" of Eurasia and not the Heartland was crucial. Others saw North America as an emerging global core, particularly with the advent of air power (e.g., de Seversky 1950). During World War II there were many calls for the development of an "American" geopolitics to challenge the German variety, pointing to the strategic advantages of an "insular" North America. Karl Wittfogel (1929; 1985), a German Marxist and author of *Oriental Despotism* (1957), criticized geopolitics for ignoring the economic context of Great Power politics and missing the real significance of "natural factors" in determining the character of political-economic development. But his goal was to prepare a Marxist geopolitics as an alternative to the nationalist varieties. After World War II a much more influential proponent of geopolitics from an American perspective was Edmund Walsh. He combined his role as a trainer of future American diplomats at the Jesuit-run Catholic Georgetown University in Washington, DC, with a powerful antipathy to the Russian Revolution and to the "global march of world communism," as he saw it. Interpreting the conflict between the United States and the Soviet Union in religious apocalyptic terms, Walsh strove to convince American policy makers and diplomats that this was no mere conflict of secular powers but "a struggle between the two great moral opposites" (quoted in Gallagher 1962, 142; see also Ó Tuathail 2000b).

To others, however, geopolitics was a vicious mutation of political geography as such. In France, the many students of Vidal de la Blache were particularly hostile (G. Parker 2001). They shared the conviction that states could not be studied in isolation from the rest of the phenomena of human geography (cities, agriculture, trade, etc.). They also saw states as reflective of the nations they claimed to represent and subject to changes in purpose and activity as population movements, trade, and other developments challenged and eroded established boundaries between states. Physical features are likewise subject to change in their roles. Albert Demangeon (Demangeon and Febvre 1935), for example, argued that for most of its history the Rhine had been a force of unity more than an international frontier. To Yves-Marie Goblet (1934) the emphasis on change and fluidity in the world political map meant that it is erroneous to treat existing states as if they are permanent fixtures without paying attention to the forces of interdependence bringing about changes in the alignment of nations and states that everyone was aware of but failed to theorize. The diversity of political territorial arrangements both historically and geographically rather than an ideal-type statehood thus needed to be center stage in political geography. Geopolitics denied this variety and justified territorial aggrandizement by large states at the expense of small ones. To one disciple of Vidal, Jacques Ancel (1936), however, the term *geopolitics* should be expropriated for other use. To him it was a branch of political geography rather than something distinctive and separate: "external political geography." He took issue with the idea of opposing political geography to geopolitics, arguing that French political geography was deficient in emphasizing the internal structure of states at the expense of examining colonialism, Great Power competition, and the geographical organization of international politics. In many respects Ancel's argument is one that would be accepted by most political geographers today. But at the time he wrote, he was very much on his own, except for similar arguments, in a different register and without recourse to labels, made by Owen Lattimore (e.g., 1940) in the United States, a specialist on China and China's inner Asian frontier.

In the United States, Bowman (1942) made great play of distinguishing his "neutral" and "objective" political geography from "crudely partisan" geopolitics. Yet his own political geography was strongly committed to the "partisan" cause of the United States. Bowman and many other university geographers were actively involved in the American war effort (Kirby 1994). But the rhetoric of science as the basis to policy prescription was vital to Bowman's career and to his image of geography within the American university. Geopolitics, therefore, had to be denied. Yet in so doing the idea that political geography could have anything systematic to say about world politics was also denied simply because Bowman's political geography had no scope for dealing with the subject matter that geopolitics, however problematically and inadequately, did try to address. Ancel's alternative geopolitics was a possibility. Another was the historicist approach to political geography pioneered by Whittlesey. These were anathema to Bowman, however, because they departed from the naturalized claim to knowledge upon which his authority as a "scientist" rested. Sadly, that was, of course, precisely the authority claimed by Haushofer and his cohorts in Germany. As David Livingstone aptly makes the point, "The territorial border between science and politics was one boundary Bowman just could not map" (1992, 253).

GEOGRAPHERS AT WAR

Despite their mobilization for the war effort around the world in World War II, seldom did geographers of whatever sort succeed in having a publishable impact on its conduct during the war or on its politics in the aftermath. Two exceptions are those of Mackinder and especially of Bowman, due to the latter's reputation as former advisor to Wilson during World War I and during the Paris Peace Conference. Not only was Bowman active in trying to dismantle the tenets of German *Geopolitik* after his article of 1942. But once the military historian Edward Mead Earle reprinted in that same year Mackinder's 1919 *Democratic Ideals and Reality*, Bowman had the idea of asking the retired British geographer to consider if his heartland strategic concept had lost significance under the new conditions of modern warfare and in view of the future peace treaties. In an article published in 1943 in *Foreign Affairs*, Mackinder (1943) claimed that the concept was even more useful than it had been twenty or forty years before (see figure 3.3). By continuing to underline the importance of the central location of Russia in Eurasia, he not only foresaw its future hegemonic position but also drew attention to how a postwar alliance between the U.S., Britain, and France could prevent any future German attack in the West, while Russia could prevent it from the East. In fact, these powers collectively did take over Germany and Berlin in 1945, even if, once the Cold War was under way, the possibility of cooperation between the U.S. and the USSR was never going to live up to Mackinder's hopes as expressed in 1943 at the height of the war. Nevertheless, that article could have contributed something to influencing the concept of a North Atlantic strategic alliance, of which, after 1949, NATO was going to be the practical translation.

Meanwhile, "Roosevelt's geographer" Bowman continued to work throughout the war for the State Department as well as in top-secret governmental consulting and postwar planning on a number of key issues: the disposition of Germany, whose dismemberment he unsuccessfully opposed, and the decolonization of the British and French Empires, where, despite differences, he shared President Roosevelt's view of their economic importance for what Neil Smith has dubbed ironically the "American *lebensraum*" (2003, 371). In fact, Smith has interpreted Bowman's main contribution as a political geographer to the two world wars, several U.S. presidencies, and peace treaties as defining the "prelude" to American globalization, an approach to world power more economic than territorial in nature. The jewel of his wartime work according to Smith (2003, 319) "was undoubtedly the establishment of the United Nations, broadly seen as a second chance at international political organization following the scuttling of the League of Nations both in Europe and in the U.S. Senate."

If Mackinder and Bowman were two exceptionally successful geographers in terms of the postwar impact of their ideas and work, less exceptional but still important was the contribution of geographers to the war, given that every country at war with a large military establishment had substantial geographical "services." Still, this was obviously different in each country. In the Soviet Union, geographers were fully mobilized for war only after 1941, under the direction of the Institute of Geography of the Academy of Sciences of the USSR and in cooperation with the ancient Russian Geographical Society. Their contribution to the war effort was dual: on the one side they provided geographic knowledge on the areas where the Red Army was fighting, including countries such as Romania, Hungary, Germany, Norway, Finland, Turkey, Afghanistan, and Chinese Turkestan; on the other side "the traditional Russian defense in depth demanded that all resources in the rear be put to use in the fullest and most rational way" (Gottmann 1946b, 161). Perhaps the most remarkable contribution of Soviet geographers was the study of how to mobilize economic resources in the Urals, Siberia, and especially Kazakhstan for the war effort (Hooson 1964). In the latter case, an enormous mapping effort became the base for the largest experiment in regional planning ever tried at the time.

France and the United States present different stories about wartime service by geographers that are worth discussing in somewhat more detail. In France, the 1939–1945 period saw war, defeat, Nazi occupation, and liberation. Between September 1939 and June 1940, a scientific group worked at the Geographical Section of the French General Staff under the direction of Cholley. After the defeat, the intense cartographic activity planned by the Service Géographique de l'Armée Française was suspended and the same office was demilitarized and renamed the Institut National Géographique. Most new maps were then produced by the African Army Map Service, which still continued field work for some years, especially in the Sahara. During the occupation, and with the country broken into administrative parts, two generations of French geographers paid a heavy price during the war, especially considering that "the traditional rivalry between the French and the German schools exposed them to more than the average 'attention'" (Gottmann 1946a, 80). Among the first generation of Vidalians, Demangeon and Sion died right after the defeat. Many were arrested by the Gestapo and sent to jail or concentration camps: Jacques Ancel died after detention in Drancy, while Musset survived two years spent in the infamous Buchenwald. Among the "grandchildren" of Vidal, some led the resistance movement, like Pierre Gourou of Bordeaux, vice chairman of the regional Committee of Liberation, while others escaped arrest, like Gottmann, who fled to the United States, where he joined De Gaulle's *La France Libre* and worked for U.S. intelligence. After the Allied landings, he returned to Paris, while from Algiers other geographers also took an active part in the war effort. During the occupation, political geography obviously lagged behind. To catch up with developments that had occurred in the main belligerent countries, a chair in political geography was established at the Sorbonne in 1945 to relaunch the subfield. In fact, during the war the only chair in political and economic geography in Paris was at the Collège de France, held by André Siegfried, who interrupted his series of studies on Anglo-Saxon democracies to publish in 1943 a human and political geography of the Mediterranean.

Indeed, despite all the difficulties, the war did not interrupt French scientific production, expressing new trends in regional studies, with an emphasis on French colonies, from the Maghreb to Indochina, as well as on other countries' colonies. On more theoretical grounds, French geography was drawing closer to biology and sociology. The need for a "social geography" became common among French geographers by 1945. Political geography was eclipsed. Gottmann, who bridged the gap between the U.S. and France, thought it was echoing developments in the U.S., where Bowman was writing, "Every geographer must be trained to understand the structure of such human societies, and how they work, for the social and economic system of a nation expresses the knowledge, idealism, standards, and dynamism of its people. It is in fact the chief national power plant" (Bowman 1945, 214).

U.S. geographers obviously enjoyed a quite different situation compared to their French colleagues, not just because scholars such as Bowman could advise the Department of State and the president on a number of issues. At the time, just about every academic field was contributing to the war effort, and geography proved especially active. By the summer of 1943, 224 disciplinary specialists of the field were employed full time in the capital, and a number part time, to the point that it was then said there were more geographers in Washington, DC, than ever gathered in any one city. According to Chauncy D. Harris (1997), their skills in providing place-specific information was especially valued in the Office for Strategic Services (OSS), the War Department, the Foreign Economic Administration, the Board on Geographic Names, and the Department of State. More than half of them worked in intelligence, gathering data or analyzing photos and maps. And most came from the University of Chicago and Clark University, but Harvard and the Universities of California, Michigan, and Wisconsin were also important contributors. Richard Hartshorne, from the University of Wisconsin, was the head of a group of seventy-seven geographers at OSS's Research and Analysis Branch, fifty-five in the Military Intelligence Division of the War Department, seventeen at the U.S. Board on Geographic Names, twelve at the Foreign Economic Administration (formerly Bureau of Economic Warfare), and twelve at the U.S. Department of State's Office of the

Geographer. The latter, created in 1921, became an independent office just two months before Pearl Harbor after a number of internal reorganizations. Since 1924 it had been directed by Samuel Whittemore Boggs, who developed and shaped it for thirty years. While in 1940 Bowman advised the president on the eastern limit of the Western Hemisphere to be the meridian of twenty-five degrees west longitude, the office was advising Congress to include Greenland and the Aleutian Islands in the same hemisphere for purposes of the Monroe Doctrine, pushing its westernmost limit to the International Date Line, after the U.S. entered the war.

As a result of these applied activities, the importance of geographic information for the practice of foreign policy analysis brought new funding to it. By 1944 the Office of the Geographer employed a staff of eighty-eight and had developed strong cartographic capabilities (Dillon 2005). Many cartographers also worked for the Military Intelligence Division at the War Department, while thirty-eight cartographers were in the OSS's Map Division under Arthur H. Robinson. The Eurafrican Division gathered data on North Africa, Italy, and other parts of Europe on a wide range of topics, from ports and transport to terrain, climate, manufacturing, supplies, and urban studies, to prepare for the liberation. Gottmann, with his French background, consulted in Washington among many others on topics such as the role of Vichy France in relation to Nazi Germany; the geography of Morocco, Lebanon, and Syria; and the strategic routes of the Sahara Desert.

According to Harris (1997), the war gave a new impulse to the field of cartography by the creation of new geography departments and map libraries, and it reinforced the professional organization of geography and increased representation of the field in the government. It also had a major impact on the geographers' careers and interests, thanks to the continuous interactions and exchanges between colleagues, to the importance of teamwork with a common goal, to the required multidisciplinarity, and to the severe self-discipline and necessity to comply with strict deadlines. In some cases it opened unexpected paths: David Lowenthal (2005) believes his interest in the geography of landscape stems from his experience in 1945 in France within OSS's Intelligence Photographic Documentation Project, a detailed terrain recognition effort aimed at landscape and human settlements. What the wartime experience did not do was provide much if any new intellectual basis for a political geography that could strike out in new directions from the naturalized conception of knowledge with which it had become fatefully connected since the late nineteenth century. If anything, the intense nationalism that the war both revealed and intensified made for greater difficulty in transcending the "my nation right or wrong" mentality by which the field had long been bedeviled.

CONCLUSION

Ratzel and Mackinder are a looming presence across the first fifty years of modern political geography. They and their major critics constitute the historic canon upon which more recent political geography has built and against which it has often revolted. Indeed, enormous effort is still put into examining their careers and ideas, often to the detriment, one could say, of moving beyond them. There are obvious discontinuities and differences across the writings of the various figures we have identified in this chapter. There are, however, a number of important continuities that also provide an important backdrop to what was to happen in political geography after 1945.

The first is the rather unrelenting focus on territorial states as the geographical units par excellence of political geography. Though not without its critics (such as Reclus), the enterprise was largely state- and empire-oriented almost to the exclusion of political processes operating at other geographical scales and in other ways. The conception of the political was almost entirely statist, with weaker liberal currents eddying around the edges. The "hard-nosed," masculinist, and realist conception of the world is here rooted in the geographical facts of an earth that rewards only those who take what they can.

The second is the merging of a naturalized claim to knowledge, based on various mixes of the "view from nowhere" and biological metaphors, with the idealist goal of serving one's own nation-state. Perhaps only Reclus and Wittfogel stand out as major examples to the contrary, at least insofar as serving a particular state is concerned. Although, by the 1930s, figures such as Owen Lattimore in the United States and Jacques Ancel in France can be seen as offering much more critical perspectives on the conventional wisdom—Lattimore because he was largely self-taught and brought his intimate familiarity with Central Asia to bear in all that he wrote, Ancel because he hoped to counter Nazi geopolitics with a more "open" version of geopolitics of his own.

The third is a "problem solving" orientation that animated all of the major figures. They aspired to influence policy, to whisper in the ear of the Prince, to paraphrase the Renaissance-era Florentine political philosopher and diplomat Machiavelli. Though geared toward establishing a presence within the new universities of the time, this goal could be served only by appearing "useful" to *raison d'état*. The prospects for political geography were thus tied to the national flag.

The fourth continuity, and one that distances us today so much from the thinking of the time, was the ready acceptance of the language of racial difference and the environmental causes of racial divisions. There was no sense of the social construction of racial differences. The American authors, often so critical of European colonialism for its subjugation of other peoples, nevertheless had blind eyes for the reality of American racism at home and in its colonies (such as the Philippines).

The fifth, and final, is that Europe and, to a certain extent, the United States are seen as being at the center of the world. The rest of the world is seen as ancillary, bit players or pawns in a world politics driven almost entirely by Great Power competition and interimperial rivalries. Of course, the world wars of the period did begin in Europe. But the rest of the world had long figured into the machinations of the powerful, and increasingly so. And it did this not just as a passive object of desire for the powerful but as an active participant in both its victimization and its own incipient liberation.

These continuities were not easily transcended. Indeed, elements of all of them live on today in political geography and beyond: an urge to naturalize knowledge claims to grasp the mantle of science, a Euro-American centered view of the world in which all others are seen as backward when compared to a Euro-American modernity, and the aspiration to influence national policies to underwrite disciplinary success and gain the ear of political leaders. But as the next chapter tries to show, much also changed after World War II.

4

Reinventing Political Geography

The ending to World War II was very different from that of the previous global conflict, World War I. At the 1919 Paris Peace Conference, geographers from Britain, France, the U.S., Yugoslavia, and other countries had been invited to advise the Big Four on territorial matters. Eight thousand kilometers of new European borders were to be drawn to partition three vanquished empires, and political geographers, together with historians and other specialists, were the natural choice. Given the accent on national self-determination, they had to help deciding on the best matches between the new territories and the actual distribution of ethnicities (Muscarà 2005a). The national scientific teams supported the geopolitical negotiations of the Big Four leaders (Britain, France, the U.S., and Italy), who planned the creation of a cordon sanitaire (see figure 3.1) to avoid future territorial expansion by Germany or revolutionary Russia through a number of buffer states (a restored Poland, a new Czechoslovakia, an Austria split from Hungary, a new Yugoslavia, a Greater Romania, etc.). This proved especially problematic wherever the new borders ran across ethnically mixed regions (e.g., in the Balkans). Geopolitics changed the formal nationality of millions of Europeans, in many cases causing migrations to match ethnicities with the new political map. A number of nations were denied their claims to independence (Ukrainians, Armenians, Kurds, Jews, etc.). Once the U.S. President Woodrow Wilson met with the consummate shrewdness and national ambitions of European politicians, the U.S. delegation's original claim to be constructing a "scientific peace" turned out to be an illusion. Despite the disputable outcomes of East Prussia and the "Danzig corridor," a restored Poland was one of the successes ascribable to Isaiah Bowman, Wilson's geographical advisor. After this "reality bath," Bowman returned from Paris "a political animal whose science was one of his major assets" (N. Smith 2003, 69). Though this made Bowman a formidable figure within American academia and as an advisor to the U.S. government in the 1930s, it led to a similar tension between the scientific and political aspects of political geography that bedeviled the more obviously politicized German version.

In 1945, however, and after the death of his sponsor President Franklin Roosevelt, neither Bowman nor any other geographer's chances of affecting decisions about the postwar geopolitical order were very good. The fragmentation of geographers' wartime work into a myriad of specialized tasks, in the context of a generational change, had diminished their ability to work on a global scale. Even at the national scale, "area experts" with country-specific knowledge but generally lacking any general geographical background increasingly took over their role. More importantly, after 1945 border making lost most of the importance it had had in 1919. Many new boundaries were to depend on the actual military deployments on the terrain at the end of the conflict and in light of the Yalta Agreement of 1944 between the United States, the Soviet Union, and Britain. The Soviet Union claimed its own cordon sanitaire over most Eastern European states, and nuclear weapons soon affected the practice of geopolitics at the global scale. In the academic sphere and among political leaders, contempt for the ideas of *Geopolitik* damaged the overall image of political geography. And even within it, the denial of geopolitics by political geographers such as Bowman was to prove doubly problematic. Not only did it sever them from considering "external political geography" and thus offering advice about foreign relations to their home or other governments, however distant that advice might be from the designs of Haushofer, it also led to a renewed disavowal of the "political" in political geography. Theory, the Germans had proved, was dangerous. Associating theory with "speculation" and the subsequent politicization of the field, political geography thus rapidly sank into intellectual irrelevance and political obscurity.

FADE OUT

As the United States took on a global political role, therefore, American political geographers refused the opportunity to either actively participate in the expanded horizons or provide the intellectual resources to review and critique what rapidly became the American popular wisdom in world affairs. Elsewhere was no better, if not worse. In the Soviet Union political geography did not formally exist, even as the political leaders and the higher echelons of the Communist Party in both subduing unrest within the country and stamping out dissent in its new sphere of influence in Eastern Europe practiced a "practical" political geography. In much of Western Europe the concern with reestablishing geography as a university subject led to both disavowal of its association with international politics and, in the context of the reconstruction from the devastations of the war, a focus on the "problem solving" role of the discipline in relation to regional and urban "problems" that could be addressed technocratically rather than politically. Still, the political tenor of the times played an important part in the eclipse of political geography. Between 1945 and 1949 a whole new geopolitical order was constructed based on the values, myths, catchwords, and political-economic orientations of the two dominant states: the United States and the Soviet Union. The intellectual genealogy of political geography was ill suited to this new world. For one thing, in their rhetoric both sides offered different conceptions of the political, on the American side liberal and on the Soviet side Marxist, from the statist one that had long dominated political geography. For another, the entire world was drawn into the bipolar conflict in a way that the world had never been divided previously. But this was seen by all sides as an ideological more than a territorial struggle, even if it had obvious geographical correlates (the "Iron Curtain" running through Europe, the threefold division of the world into the U.S. and allies, the Soviet Union and allies, and a "Third World" of mainly former colonized countries in which the U.S. and Soviet Union competed for influence, etc.). Of course, a more adaptive field in touch with intellectual currents in adjacent fields such as political science and diplomatic history might have had something of a more positive response. The personal danger of questioning the conventional wisdom needs emphasizing. In the United States, figures who did offer alternative perspectives were likely to receive subpoenas to appear before congressional committees investigating "un-American activities" or find difficulty in acquiring academic posts. As it was, in the 1950s nothing very new emerged in political geography in

Western Europe and North America that was not based on the conventional wisdom established by Bowman and a small group of fellow travelers, of whom Richard Hartshorne (1950) and Stephen Jones (1954) were perhaps the most influential.

The revitalization of geography as a whole in the United States in the 1960s had knock-on effects in political geography. A new generation of geographers discovered an interest in the spacing of social forms, such as urban settlements, land uses, and migrant flows. They increasingly addressed these empirically using the quantitative research methods popular in fields such as economics, demography, and urban sociology. The net effect was to improve the social status of geography as a university subject, at least among adjacent fields in the social sciences. Though the initial revival was to become the subject of controversy, particularly over theoretical and methodological issues, it set the scene for a revival of political geography. Increasing opposition to Cold War sensibilities was also very important to the revival. While the nuclear arms race gathered pace and Cold War hysteria entered one of its darkest moments, a new generation started to react democratically by beginning to ask how the Cold War had arisen and how it could be made less dangerous. The paradoxical atmosphere of the time is well caught in Stanley Kubrick's film *Dr. Strangelove,* which in exposing the continuity between World War II and the Cold War also underlined the tragic risk of an accidental nuclear holocaust by presenting it as a comedy.

As a post–World War II generation was coming of age, the civil rights struggle in the United States and the Vietnam War were key in opening up discussion about the assumptions upon which the Cold War had been based. They drew attention to the contradictions between what the United States stood for in Cold War discourse and the reality, as many people saw it, in the United States and in U.S. government behavior around the world. How true to its self-confessed beliefs about human rights was the United States? Was not the Vietnam War the outcome of a civil war rather than an instance of the global conflict between "democracy" and "communism"? In a wide range of fields the Cold War had had intellectually stultifying effects. It had encouraged the idea of a permanently divided globe in which idealized ideological differences between the United States and the Soviet Union were all that mattered. This was shattered in the 1960s. In our opinion, it is no exaggeration to speak of a reinvention more than simply a revival of political geography as an older generation passed away and was replaced increasingly by a more diverse and intellectually adventurous group of academics coming to intellectual maturity at a time of great social and political change.

As a result, research in political geography began to revive, connecting the field both to other currents within geography and to relevant work in other disciplines. It could be useful to think of the reinvention in terms of three intellectual "waves" sweeping through the field between the 1960s and 1990s: from the spatial-analytic focus of the earlier years (and onward, because each wave has continued to flow), followed by radical political-economic perspectives in the 1970s and 1980s, and then by postmodern perspectives from the late 1980s.

OVERVIEW

In this chapter we want to do several things. First of all, we wish to provide some sense of the geopolitical context that political geography had to face after World War II and, with a few exceptions, the irony of having little or anything new to offer in the way of understanding it. Then we want to account for the reinvention of political geography beginning in the 1960s. This was not a "one-shot deal" but a cumulative series of influences from both inside and outside the world of universities over the period between the 1960s and the 1990s. Much of the chapter, however, is taken up with laying out and providing examples of the three intellectual "waves" that have swept through the field and still provide its basic intellectual structure today. This involves an extended discussion of the five major substantive areas of research in political geography (geopolitics, the spatiality of states, geographies of social and political movements, places and identities, nationalism and ethnic conflicts) from the three different theoretical perspectives. An example of how the three theoretical perspectives would approach the same subject is provided at the outset of this section, with the empirical case of the political geography of the 2008 U.S. presidential election, to provide a sense of their fundamental differences with respect to a common substantive topic. We end with a discussion of how the "edges" between the three sets of perspectives have started to erode as new perspectives trying to draw from more than one are beginning to emerge. This leads the way into the next chapter, where the post–Cold War geopolitical context is seen as profoundly affecting both theoretical perspectives and topics of research.

THE COLD WAR GEOPOLITICAL CONTEXT

The total victory of the American-British-Soviet alliance over Nazi Germany and Imperial Japan and the deployment of forces it produced in 1945 had two immediate consequences. One was that Soviet influence now extended over Eastern Europe and into Germany. This stimulated both a direct confrontation between the U.S. and the Soviet Union and a continuing U.S. military presence in Europe to "contain" possible Soviet expansion. The U.S., concerned about revitalizing world trade and American economic development, needed to protect and aid potential allies in order to meet its goals. The other consequence was that there was little major opposition in the U.S. to a "forward" U.S. presence in Europe and, increasingly, elsewhere around the world. Unlike after World War I, when the United States turned its back on a major global role, this time there seemed little alternative. Europe and Japan were physically, economically, and psychologically devastated. What domestic opposition there was, on both left and right, rapidly disappeared after 1947 with the Soviet subversion of Czech democracy. The view that Greece, Italy, and France, in different ways, faced the prospect of possible Soviet-leaning governments, the Soviet test of its first nuclear weapon in 1949, and the "loss" of China a month later to communist revolution all added to the existing fear that political and economic chaos abroad would have extremely negative effects on the United States itself.

Such activities as the Marshall Plan for European economic recovery, the founding of NATO to coordinate military planning between Western Europe and North America, and the creation of such organizations as the General Agreement on Tariffs and Trade (GATT) and those emanating from the Bretton Woods Agreement of 1944 (in particular the IMF and the World Bank) represented major U.S. initiatives to incorporate Western Europe into the U.S. sphere of influence, combat Soviet influence around the world, and make the world safe for American business enterprise. The overall logic of the American approach was that military expenditures would provide a protective shield for increased trade across international boundaries. This would, in turn, redound to American advantage. Making this possible, however, required establishing globally those institutions and practices that had already developed in the United States, such as mass production/consumption in industrial organization, electoral democracy, limited state welfare policies, and government policies geared toward indirect stimulation of private economic activities. Taken as a whole, these features constitute an "embedded liberalism" that American leaders believed would provide both economic growth in Western Europe and around the world and protect the United States from the political-economic threat posed by the Soviet Union.

It is clear, however, that at the end of World War II the two superpowers faced very different economic conditions. Despite all the concern for its economic future, during the conflict the U.S. had overcome the Great Depression and its national territory had suffered no damage. The Soviet Union had suffered the largest population losses of any of the combatants, and its economy was almost entirely organized for military production. Its dependence on centralized planning meant that it was not designed very well for production of consumer goods. If in four years of war U.S. industrial production had surpassed the prewar combined value of all the other industrial powers, whose countries and economies were now ruined, it proved relatively easy to convert American industrial power into civilian production. For the "Arsenal of Democracy," shrinking production was never an option, for if on the one hand increasing unemployment opened up the country once more as it had in the 1930s to the risk of social unrest, on the other exports and foreign investment offered the prospect of building markets for U.S. businesses well beyond American shores.

As in most of Europe, in the Soviet Union the war had caused immense devastation and massive casualties. Nevertheless, despite all the war destruction, under the dictatorship of Stalin from 1943 to 1947 the Soviet Union constructed a formidable military economy that required as its premise the existence of a major external threat. The recent experience of invasion by Nazi Germany meant that the idea of external danger was not hard to sell to the Soviet population. Associated with this sense of external threat was a popular identification with regimes and revolutionary movements that were opposed to the reinstatement of capitalism-as-usual around the world. Within the Soviet bloc, the fear of the United States was used to justify an incredibly brutal repression of political dissent through a vast system of concentration camps (the so-called *gulag*). "Enemies of the people" could be sent away on the flimsiest of grounds, condemned using biological-hygienic terms not dissimilar to those applied by the Nazis to Jews and other "racial inferiors": vermin, pollution, and "poisonous weeds" requiring "ongoing purification" through peremptory imprisonment for long terms. Thus the Cold War had two sets of roots: one set in the United States in the fear of a repeat of the Great Depression of the 1930s if international commerce was not rapidly reestablished, and the other set in the Soviet Union with the fear that the Soviet experiment in a planned territorial economy and society would fail if it did not rise to the challenge posed by American containment.

By European standards, both the United States and the Soviet Union were peculiar states. They both had origins in revolutions with explicit ideological agendas. They both claimed popular mandates that transcended particular ethnic, class, or even national interests. They offered themselves as uplifting examples of political-economic experimentation in a world mired in postwar poverty and gloom. Within the two countries, the lack of social and ethnic homogeneity meant that precisely what was either "American" or "Soviet" was unclear, so the threat of the "un-American" and the "anti-Soviet" became central to official definitions of national identity. In each case, the threat of external danger from an equivalent but mirror-image superpower anchored national populations, and intellectuals, into a political consensus about the broad parameters of "national security" (Robin 2001).

The main consequence of this shared sense of vulnerability was an idealization of each by the other. Each became a super potent adversary in the eyes of the other. As the United States came to personify capitalism to

Soviet citizens, so did the Soviet Union represent communism to Americans. Each became the geographical manifestation of a totally opposing political economy. Each was also seen by the other as uniquely powerful: a real threat without the human flaws and institutional drawbacks that each saw in itself. Definite domestic interests were served by this geopolitical reductionism. In the Soviet Union it disciplined potential dissidents into supporting a monolithic state apparatus. In the United States it produced a consensus around economic policies expanding mass consumption, a permanently enlarged military (with a budget to match), and opposition to any politics (usually of the left) that could be construed (through guilt by association with the Soviet Union) as subverting the United States from within. In short, it eroded American democracy. The very identity of being American became associated with a narrow political spectrum at home and a virulently antisocialist/anti-Soviet position (they were usually not distinguished) abroad.

Crucially, however—and this clearly sets the United States apart from the Soviet Union—the process of representation in the United States also offered the possibility of opening up American politics to those whom it had hitherto largely excluded from public life when they challenged the reality of the claims made on their behalf. Thus the black civil rights movement and movements to expand the rights of a wide range of groups, from women to gays and lesbians, could point to the U.S. Constitution and Bill of Rights as mandating equal treatment throughout the United States, if the United States was indeed the "homeland of freedom" it claimed to be in U.S. Cold War discourse.

In the geopolitical setting that arose from the late 1940s onward, the ideology of the Cold War developed the following major characteristics: a central systemic-ideological conflict over political-economic organization; "three worlds" of development in which the American and Soviet spheres of influence (respectively, the First and Second Worlds) vied for expansion in the Third World of former colonies and nonaligned states; a homogenization of global space into "friendly" and "threatening" blocs in which idealized models of democratic capitalism and communism reigned free of geographical contingency (over there is all like that); and the naturalization of the ideological conflict by such key geopolitical terms as *containment*, *domino effects* (linking distant events back home through the image of falling dominos), and *hegemonic stability* (each side needs a leader to enforce discipline on the others).

It is little exaggeration to claim that in the five decades after 1945 American influence was at the center of a remarkable explosion in what can be called "interactional" capitalism, moving beyond the territorialized approach hitherto dominant in the world economy and as evidenced by the interimperial rivalry of the period 1875–1945. Based initially on expanding mass consumption in the industrialized countries of Europe, North America, and Japan, it later involved the reorganization of the world economy around a massive increase in trade in manufactured goods and foreign direct investment. Abandoning territorial imperialism, "Western capitalism . . . resolved the old problem of overproduction, thus removing what Lenin believed was the major incentive for imperialism and war" (Calleo 1987, 147). The driving force was mass consumption in the industrialized world, particularly in the United States. The products of such industries as real estate, household and electrical goods, automobiles, food processing, and mass entertainment were all consumed by increasing numbers of people within (and, progressively, between) the producing countries. The welfare state helped sustain demand through the redistribution of incomes and purchasing power. If in the late nineteenth and early twentieth centuries the prosperity of the industrial countries depended on favorable terms of trade with the underdeveloped world of Asia, Africa, and Latin America, demand was now stimulated at home. Moreover, until the 1970s the income terms of trade of most raw materials and foodstuffs tended to decline. This meant that the prices for such goods went down even as the cost of manufactured imports increased. This had negative effects in general in the Third World, but it encouraged some countries to switch to export-based industrialization that later paid off as they found lucrative export markets for their manufactured goods (as in, e.g., Taiwan, Mexico, South Korea, and China). The globalization of production that followed has slowly undermined the clear identity between products and the places they are manufactured as different phases in production are located in different countries depending on what mixtures of costs and benefits they offer to producers. It is not clear how sustainable a world economy can be that involves a fundamental split between the location of production, particularly of manufactured goods, on the one hand, and the location of consumption, on the other.

A vital factor in allowing the United States a commanding role in the world economy, even as its own economy often sputtered rather than totally outproduced others, was the persisting but historically episodic political-military conflict with the Soviet Union. This had two major peaks in intensity in the late 1940s and early 1980s when each side perceived the other to be increasingly hostile and dangerous and in the early 1960s when the U.S. government worried about Soviet advantages in satellite and missile technologies. The mid-1970s was a period of maximum cooperation or détente when American goals of retreating from the military disaster of Vietnam and cutting military spending coincided with Soviet interest in stabilizing military expenditures. Each side blundered into major military misadventures that came to have major domestic consequences. For the U.S. it was Vietnam, where a political commitment to the government of South Vietnam in the late 1950s led to a massive buildup of troops by the late 1960s but with increasing opposition from within the population of the United States, who were never clear what the war was about. Claims that the war was to prevent the spread of communism seemed implausible to those, including not a few who had spent time fighting in Vietnam, who saw it as a civil war between competing factions rather than a local manifestation of the Great Global Struggle between Good (the U.S.) and Evil (the Soviet Union). For the Soviet Union it was the intervention in Afghanistan in 1979 to prop up an allied government in the face of hostility from religious and ethnic opponents who received backing from the United States through Pakistan. In this case military failure was also compounded by popular distaste for the intervention in the Soviet Union as largely unrelated to either Soviet national security or to the support of noble ideals.

Even in periods of détente, the overarching Cold War served to divide the world into two spheres of influence and to tie allies into this geopolitical structure (see figure 4.1). For a long time this imposed an overall stability on the world, since the U.S. and the Soviet Union were the two major nuclear powers, even as it promoted numerous "limited wars" in the Third World where each of the superpowers intervened or armed surrogates to prevent the other from achieving a successful "conversion." For all their material weakness, however, Third World countries—which since the end of World War II had multiplied as the withdrawal of European powers pushed forward decolonization—had considerable leverage. They had to be wooed, and often they resisted. The world map was no longer a "vacuum" to be filled by the Great Powers. The global military impasse between the superpowers protected the territorial integrity of existing states. Any disturbance of the status quo threatened to bring down the entire structure on the heads of all concerned.

Few expected this system ever to change. It became a part of everyday life around the globe. In the end it was the collapse of the Soviet Union that brought about the demise of the Cold War. Undoubtedly, many factors precipitated the Soviet collapse between 1989 and 1992. Among these pride of place must be given to the following: the failure of the Soviet economic system to provide a rising standard of living to most of the population, an increasing technological lag behind the rest of the developed world (its rigid internal organization could compete somewhat in an industrial economy but not in a postindustrial information economy), the burden of a huge military budget, and the disappearance of almost any political idealism from the political elite. The final page in the history of the Soviet Union began in its sphere of influence, with social and political movements in East Germany and Lithuania challenging Moscow's central authority on the basis of deprival of human rights (Thomas 2005). When dissent spread without adequate central response, either repressive or reformist, the whole system suddenly collapsed. The lack of accurate information about Soviet

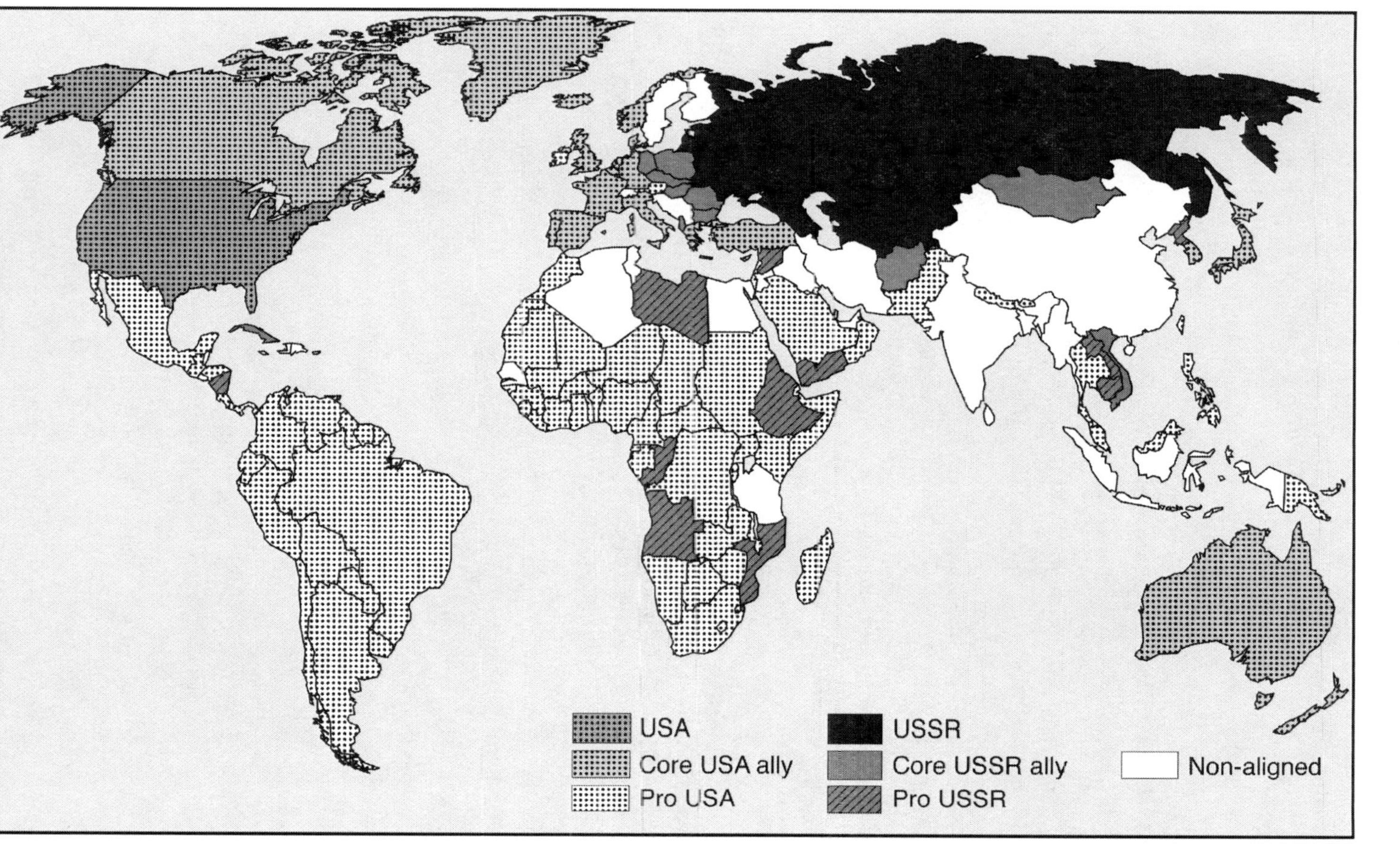

Figure 4.1. The Cold War's bipolar order divided the world into two main "spheres of influence," one centered on the U.S. and one on the USSR, as of 1982.

Source: Authors

politics and the taken-for-granted quality of the ideological confrontation at the heart of the Cold War prevented Western commentators from recognizing the Soviet system's multiple failures (Hooson 1966, 1984; Sidaway 2009). As a result, on the western side of the Iron Curtain, the sudden collapse was absolutely unexpected. Its impact worldwide was huge. Not only did the certainty and stability that the Cold War gave to world politics disappear, but this coincided with an increased global economic interdependence that introduced added uncertainty and a lack of global institutional means to deal with it. As the Soviet Union disintegrated, its people had to define themselves anew in largely ethnic-national terms, because there was little else upon which to rebuild political identity.

The United States also lacked an external enemy against which to define national identity following the Soviet collapse, at least until the devastating terrorist attacks on targets in New York City and Washington, DC, on September 11, 2001, provided a possible alternative. In the next chapter, we will see how shadowy networks of terrorists devoted to this or that cause but united in their hatred of the United States as a symbol of world capitalism, globalization, intellectual freedom, or religious diversity (take your pick) for almost a decade came to be viewed as a substitute for the former Soviet Union. This was a fateful choice because arguably it has drained massive financial resources from the United States at a time when the country has faced a much more potent economic challenge from globalization and the rising position of China and others within the globalizing world economy. Be that as it may, those in U.S. governments during the Cold War who thought in terms of preemptive war against the Soviet Union to "roll back" the communist threat were finally able to realize their dream of rolling back another putative enemy in the invasion of Iraq in 2003 (Bacevich 2011). But this dream turned into a nightmare on the ground with the fiasco of an occupation that had not been planned. The more profound problem with seeing Islamist terrorism or Middle Eastern "rogue states" as a substitute for the old Soviet threat is that they offer little that is attractive to people in general. The essentially negative example offered by Islamist terrorists suggests that they will not easily substitute for what has been lost with the demise of the Soviet Union. While the Soviet Union tied allies such as Germany and Japan closely to the United States, there is not much evidence that Islamist or other terrorist groups provide much of an agenda that is attractive to many residents of most Islamic countries, let alone to U.S. allies or within the United States itself. Whatever its failings, we should never forget that the former Soviet Union represented an alternative model of modernity to the United States. That was why it was seen as so threatening. Its challenge was never just about a military threat. Satan, as the great English poet John Milton knew only too well, must be tempting as well as evil.

THE DILEMMA OF POSTWAR POLITICAL GEOGRAPHY

The dilemma facing political geography after World War II was twofold. One part was the need to move beyond the old frameworks at a time when an ideological dispute based on political-economic differences between the United States and its erstwhile ally the Soviet Union had largely displaced the fundamentally territorial interimperial rivalry of the period 1875–1945. Although attempts were made to fit the U.S.-Soviet Cold War into a Mackinder-style geopolitical model, this never matched either the technological or the ideological conditions of the Cold War. The political geographers of the time had nothing else much to offer. The political geography textbooks of the period 1945 to 1960 in the United States and Britain (e.g., Pearcy, Fifield, et al. 1948; Weigert et al. 1957; Alexander 1966) cover more or less the same ground using the same organizational frameworks as the texts from the 1930s. Political geographers defined themselves as teachers rather than researchers; an unfortunate trend when major universities, especially in the United States, were defining themselves as research universities. As political geographers failed to come to terms with new conditions, their possible roles were usurped in the United States by figures in such fields as diplomatic history and international relations who rushed in to fill the intellectual gap in providing advice and geopolitical soothsaying to demanding governments. Area studies programs staffed by nongeographers arose to fulfill the traditional role of university geography departments in teaching students about "foreign areas."

It would be wrong to focus simply on the deficiencies of the personalities involved, such as Bowman and others, or on the complacency and provincialism of many geographers. The intellectual genealogy of the field was a considerable burden. After all, the Nazi geopoliticians were part of the history of the field. To the extent that the field had intellectual forebears, their ideas were not readily translated into the context of the Cold War as it was understood in the United States and Western Europe. Still, the contemporary divorce of economic and political geography also hindered the possibility of reconstituting political geography around the workings of public sector investment or the workings of local governments in the absence of willingness to tackle international issues in other than traditional inventory terms. Above all, however, this was not a period of major theoretical innovation in any of the social sciences, except offering the blandest and most officially acceptable interpretations justified by application of the word *scientific* to frequently descriptive and uncritical studies. In the period from the late 1940s until the late 1950s, a political correctness stalked the United States in the form of McCarthyism. This was the fear that academic research and writing would be labeled as subversive and un-American. Political geographers had the example of one of their own to bear in mind every time they put pen to paper from 1950 onward: the case of Owen Lattimore.

This case broke into public view in March 1950 when Senator Joseph McCarthy (Republican of Wisconsin) named Lattimore as the "top Russian agent in this country [the United States]." At the time, McCarthy was at the height of his powers, naming former government advisors and officials, such as Lattimore, and a whole variety of others, such as Hollywood screenwriters and directors, as Communist agents and "dupes." This witch hunt descended on Lattimore because of his writings on China and close association with the Institute for Pacific Relations (IPR), an organization founded in 1932 that funded and published research and writing on the Far East, and that McCarthy and his allies contended had become a "Communist-front organization." Lattimore had a long public track record of his views on the internal dynamics of Asian countries and the geopolitical consequences of Great Power policies (e.g., Lattimore 1945, 1949). His political "mistake" lay in refusing to see the Soviet Union as a "red menace" and in trying to understand the appeal of the Communists in China. But until 1948 he had encouraged the possibility of a rapprochement between Chiang Kai-shek and the Communists in China. At the time, McCarthy and others, most significantly Senator Richard Nixon (Republican of California), were looking for those in the United States who had "lost China" to the Communists in 1949. In twelve days of grilling before the Senate Internal Security Subcommittee (from February 26 to March 21, 1952), Lattimore attempted to rebut the charges and the evidence of previous antagonistic witnesses (including Karl Wittfogel). The Report of the Subcommittee, as David Harvey relates (1983, 7), "relieved Mao and millions of Chinese of any responsibility for their revolution and concluded that 'but for the machinations of the small group that controlled [the IPR], China would be free, a bulwark against the Red hordes.' It also recommended the indictment of Lattimore for perjury, charging that he had been 'a conscious articulate instrument of the Soviet conspiracy' since the 1930s."

The charges against Lattimore were eventually dropped on June 30, 1955, but not until he had been subject to numerous court appearances and considerable legal costs (L. S. Lewis 1993). Meanwhile, he had been on leave of absence with pay from Johns Hopkins University in Baltimore. His position on campus proved difficult, however, and in 1963 Lattimore left the United States to lead a Chinese Studies Program at the University of Leeds in England. The Lattimore case illustrates a general point about the first part of the dilemma of postwar political geography in the United States:

> how dangerous it is to cultivate perspectives on geopolitical questions that deviate from certain narrow conceptions of national interest or offend some dominant political line. Hardly surprisingly, geopolitics dropped out of Ge-

> ography and political geography became a dull backwater after the McCarthy years. Geographers felt safer behind the "positivist shield" of a supposedly neutral scientific method. All of this adds up to a decided abrogation of social responsibility, an understandable but ignoble currying of professional safety and security in a highly politicized world. (Harvey 1983, 10)

The second part of political geography's postwar dilemma was the crisis of the naturalistic approach to knowing that had hitherto dominated the field. In particular, the attribution of causal powers to physical features had lived on in geography in general, and in political geography in particular, long after other fields such as economics, cultural anthropology, and psychology had abandoned it. Yet to many geographers, particularly in the United States, abandoning the essential grounding of human (and political) geography in the facts of nature was to abandon geography as a subject. Of course, the essence of the Cold War was a denial of the importance of relative location or physical features as the determinant of conflict and the claim that ideological competition had displaced the imperial rivalries of the previous epoch. Consequently, because they were out of touch with the new era, leading figures in political geography focused more on policing the boundaries of the field, declaring what was and was not geography, rather than engaging in much innovative or novel research. The political geographer Richard Hartshorne is the exemplary case. He largely abandoned research on political boundaries to excavate geography's past in German regional geography and defend the "exceptionalism" of geography as a field bridging physical and human phenomena using the idea of the "region." This doesn't mean that Hartshorne's arguments about regions could not be considered insightful and useful. Whether someone of such a conservative cast of mind as Hartshorne, however, could have ever challenged conventional Cold War thinking even if he had not buried himself in policing the boundaries of geography and devoted himself to the cause of the political boundaries he had studied in his youth is open to question.

Further reinforcing the commitment to naturalism was the attempt to reestablish geography's intellectual reputation—after the politically disastrous association with Nazism and the administrative blow of the closure of Harvard University's department after World War II (where Derwent Whittlesey had given political geography a high profile)—by showing the intellectual *bona fides* of the field. This was attained by clarifying its "nature" as defined by Germanic scholarship (in particular that of Hettner) and adopting the vocabulary of the "hard sciences," which, in contradistinction to "soft" fields such as geography, impressed grant givers in Washington, DC, and private business with arguments liberally laced with terminology such as the "fact-value distinction," "objectivity," "disinterestedness," and "the scientific method."

The American research universities of the postwar era, however, were singularly unreceptive to geography's claim to scientific status, given its seeming lack of utility in providing the state with neutral research results on which pragmatic public policies could be based. Its focus was just too broad or "generalist" and its methods just too different from those of the laboratory sciences that had the highest kudos (see, e.g., Martin-Nielsen 2010). Louis Menand (2001, 45) makes the general point: "scholarly tendencies that emphasized theoretical or empirical rigor were taken up and carried into the mainstream of academic practice; tendencies that reflected a generalist and 'belletrist' approach were pushed to the professional margins, as were tendencies whose assumptions and aims seemed political." James Bryant Conant, chemist and the president of Harvard who closed the geography department there, was also one of the main designers of the system of federal funding of universities that oriented these institutions to providing knowledge judged relevant for fighting the Cold War (Geiger 1993; Graham and Diamond 1997). If, as Ambrose Bierce once wryly observed, "war is God's way of teaching Americans geography," then the Cold War was to prove the exception to the rule.

There were some who did offer relatively new departures in political geography, suggesting that the tenor of the times was not the only factor in relegating the field to the margins of academic enterprise. Their careers were outside the mainstream academic orientations of the times, even though they did have personal and intellectual connections to mainstream figures such as Hartshorne. What was more important is that they were much more broadly educated and mixed with scholars from a wider range of backgrounds than was typical of most American and European geographers. The people we have in mind are Jean Gottmann and Harold and Margaret Sprout.

As mentioned in chapter 2, Gottmann was an iconoclastic figure. An academic and linguistic nomad, Jean Gottmann (1915–1994) moved backward and forward across the Atlantic for over twenty years. A refugee twice over, from Kharkov in the Ukraine to France in the aftermath of the Russian Revolution and from France to the United States following the Nazi invasion of France, Gottmann was the veritable citizen of the world in his work as in his life. He was the first major political geographer for whom national allegiance was not a driving force in his thought. Though he was more than sympathetic to the formation of the State of Israel, Gottmann's political geography is truly cosmopolitan. He came to speak to two audiences, largely outside the confines of academic geography itself. One was that of specialists in international relations, the other that of urban planners and architects (Muscarà 1998). Though Gottmann saw his work in political geography as closely related to that on cities, his impact was largely divided between the two communities. His lack of orientation to the world of academic geography in either France or the United States possibly explains both the lack of attention given to him in disciplinary histories and his lack of concern to ingratiate himself with the conventional wisdom in political geography. Since the 1990s, beginning in France and Italy, and increasingly so in the Anglophone sphere, there has been a renewal of theoretical interest in the political geography of Jean Gottmann that has ranged from his iconography/movement heuristics (e.g., Prévélakis 1996; Sanguin 1996; Sanguin and Prévélakis, 1996; Muscarà 1996, 2001, 2005b, 2009; Hubert 1998; Bruneau 2000; Eva 2001), the centre/periphery binary couplet (Agnew 2002), the relation between his political and urban geography (Agnew 2003b); and his understanding of territory (Johnston 2003; Muscarà 2005b; Agnew 2009; Elden 2010) to the central relevance of the notion of accessibility in his political geography (Labussière 2011).

Having started his education in political geography under Demangeon in Paris in the 1930s, at a time when French geographers were publicly criticizing German *Geopolitik* (see chapter 3), Gottmann was always well aware of the risks of environmental determinism. Only after an impressive amount of wartime-driven empirical work for the American and French governments and the UN during his twenties, however, did he finally focus on his theoretical position, which he intermittently continued to develop throughout the 1950s and in a variety of forms down until the 1970s and 1980s. Gottmann displayed a communitarian rather than a statist or liberal conception of the political and framed political geography within the tradition of Vidalian human geography, which vests an explanatory power in the role of human flows or movement (and of the crossroads they generate). In his early work (1947, 1952), he sees the political partitioning of the world as a product of both (1) interactions between forces of external change (movement/circulation) that move people, goods, information, and ideas and (2) the symbols and beliefs of territorially defined social groups (iconographies) who stabilize their existence by a "common mooring" in terrestrial space. Iconographies could include elements of the social, religious, or national history of a community (in the latter case exemplified by the national flag, national anthem, currency, stamps, borders, the capital city, and specific "sacred" sites endowed with specific meanings). Still, what really matters is not so much the actual contents of an iconography, but the action it can exert as a whole on borders and on the acceptance of foreigners, to the extreme of dictating the expulsion from the territory of all those foreigners carrying iconographies that are deemed "incompatible" with the preservation of the local (or national) identity. In this sense, he defined the domain of political geography as that of *la repartition des accès*, or the public allocation of accessibility (1966).

Gottmann found that this was already present in classical philosophy and ancient history and he connected the basic two elements in tension with the opposition between Plato's ideal city-state on the one hand, a closed, protected, and largely self-sufficient territorial entity, and the Alexandrine network of cities on the other, an open, accessible system of connected nodes. If the former represents the total victory of iconography, for the latter circulation is the prevailing force. The political geography of nation-states, therefore, need not be a permanent state of affairs. The political partitioning of the world is a product of the interplay between systems of movement and systems of resistance to movement (iconographies). The distribution of political power is thus seen in terms of dynamic tendencies rather than permanent situations (Gottmann 1952, 1973, 1980). Part of the beauty of Gottmann's approach, therefore, and one reason why it resonates so well in contemporary political geography is its emphasis on historical contingency and the refusal to see a system of nation-states as necessarily representing the ultimate form of global political organization.

Since the political behavior of people is a moral issue, deeply rooted in human psychology, as André Siegfried for one had suggested, the oscillation between the two polarities is not just related to socioeconomic imperatives but corresponds to two main psychological drivers: the opening up of territory to movement is driven by a search for opportunities, and the closing down of territory in the name of an iconography is driven by the search for security. Security and opportunity are therefore the two main functions of territory and its two main psychological drivers. Even if they usually coexist simultaneously, in specific historical and geographic conditions one tends to prevail over the other, thus affecting relative accessibility to different territories.

In times of peace, communities that do not feel threatened may be more open to opportunities beyond their borders, maximizing the relevance of movement and accessibility, allowing for more tolerance toward cultural diversity and innovative ideas. Symmetrically, when fear prevails, as in wartime, the search for security will enhance control over whom and what has the right of access to the space of a community, reaching the extreme of dictating the closure of borders to foreigners, as happened in Japan (1639–1853) during the Edo period under the Tokugawa Shogunate (see, e.g., Perrin 1979). Such closure may well be seen not just as motivated by physical security, but especially as a self-defense reaction to reduce the impact of foreign cultures on a given, in this case Japanese, identity. In fact, for over two centuries only the Dutch and the Chinese were allowed to trade, though confined to a small island built on purpose in Nagasaki (itself a harbor developed by the Portuguese in the second half of the sixteenth century). Once a world of modern nation-states became the rule, the opening and closing of national borders thus reflected not just the need for physical security but also one for the psychological self-defense of a community's identity and culture from foreign pressures, which implies also a greater attachment to those sets of symbols that are the cement of such communities.

From this perspective Gottmann came to see territory as a "psychosomatic phenomenon." Psychosomatic implies that territory is an extension perhaps not so much of the individual body (in the sense that the media theorist Marshall McLuhan saw technologies as human extensions) but of the body politic and as such reflects its changing psychology. Its function is therefore to provide a certain degree of separation as the indispensable medium for mediation between different communities (Gottmann 1973). In this construction, following World War II territory began to embody an international "social" function in providing the necessary mediation between different national communities. Increased globalization of the world economy reinforced this trend. In this light, Gottmann warned that it may not be possible to imagine the fusion of all the iconographies into a single one. Territory, therefore, still needs to provide "a certain degree of separation" between different communities in order to preserve their iconographies, to protect their differences (Gottmann 1994). In light of this mode of thinking and in the wake of the post-1945 nuclear bipolar order, Gottmann adhered to the vision that the world community should manage nuclear weapons. He viewed the world community as slowly emerging through the networks of cities rather than in terms of a consolidation of existing nation-states (Gottmann 1994).

Gottmann's *Megalopolis* (1961) is the notable example of this logic. In extending the scale of analysis from a single city to the urbanized northeastern seaboard of the U.S. as a unitary urban/regional system, he pioneered the new paradigm of future global city-regions (Agnew 2003b). In the Cold War ideological context, *Megalopolis* was praised by some and opposed by others as an ideological statement in favor of the U.S. In fact, the link between political and regional/urban geography was missing. U.S. governments did fear the potential impact of nuclear attack on U.S. cities (Farish 2010), especially where they were most concentrated as in the Northeast. In suggesting for the region's name the same "mega" prefix used to measure the destructive power of the new hydrogen bombs of the 1950s (megatons), Robert Oppenheimer, a major figure in the history of U.S. physics and nuclear weapons, for one was aware that the Boston-Washington urban corridor Gottmann studied matched the destructive scale of the new weapons. Later, he introduced Gottmann to the private foundation that financially supported the five-year study on which the book was based. Their common view of Megalopolis was beyond the technocratic. Oppenheimer thought the nuclear revolution had to be politically managed beyond the nation-state, at the international scale, and Gottmann came to see in megalopolitan growth a trend that in the long term could contribute to the birth of a new political order based on a global community of cities. "Preparing ground for a world community," he wrote in a book reflecting on the political implications of Megalopolis, "I am convinced that the world must become a community of cities before it becomes, if ever, a community of nations" (Gottmann 1994).

If Gottmann represented a clean break with much of the political geography of the 1950s, Harold and Margaret Sprout seem at first sight very much in the mainstream. Affiliated with the Center for International Relations at Princeton University, the husband and wife writing team often refer to the writings of Bowman, Hartshorne, and others in their many works published between 1939 and 1978 (Sprout and Sprout 1939, 1943, 1962, 1965, 1978). Perhaps their most important book, *The Ecological Perspective on Human Affairs* (1965) draws from a series of essays previously published in the 1950s. What is distinctive about the Sprouts' writing is their reworking of the physical-human divide in geography in terms of an "ecological" perspective that surveys the ways in which environmental factors are often seen by practitioners (and scholars) as driving the course of international politics and proposes a behavioral approach that incorporates such perceptions and the limitations that environmental factors exert on the accomplishment of policies prosecuted with perceptions of environmental effects in mind.

The distinction between (1) making decisions and (2) operational results and the different role of environmental factors—populations, resources, and others—in each are fundamental. To the Sprouts this is where environmental determinism was particularly problematic. It presupposed that political outcomes could be predicted from environmental and locational conditions. They used their framework also to criticize the psychological and system ideas current among American scholars of international relations, arguing that the lack of attention paid to the geographical milieu in which political leaders make policies reduces the likelihood that decisions will yield desired results. A "possibilism" similar to that advocated by Vidal de la Blache informs this perspective. But it is one in tune with the behavioral orientation of American social science in the 1950s and 1960s, determined to eschew statist understandings of the political in favor of a liberal one that gives political leaders and popular opinion a strong role and thus plays up "the fruitlessness of deterministic predictions" (Sprout and Sprout 1965, 199).

Gottmann and the Sprouts offered routes out of the impasse into which political geography had fallen. If Gottmann offered a political geography informed by wide reading in political theory and a personal history that saw political territoriality as always historically contingent, the Sprouts provided a way of "keeping the physical in political geography" while placing it in the context of human perception and decision making. In both cases, however, the lack of institutional connection to academic geography and their orientation to other fields (such as international relations and urban planning) that did not necessarily have a theoretical commitment to the issues they identified reduced their impact and left the mainstream of political geography much as it had been since Bowman.

WHY REINVENTION?

Geography as a whole revived in the 1960s. This had much to do with the massive expansion of higher education in North America and Western Europe and with the desire to provide schoolteachers with some knowledge of the world and lay the groundwork with all students for a more educated citizenry. But it also had much to do with developments that made the subject both more "scientific" and thus attractive to government funding agencies. Initially, political geography failed to attract much attention from the "new" geographers. They were obsessed with modeling the impact of distance on settlement spacing, flows of migrants from place to place, and the location of industries and agricultural land uses. These topics lent themselves to both the quantitative methods that were seen as the measure of the "science" the new geographers claimed as their mantle and the focus on spatial analysis that they saw as the future focus for the field as a whole. The subject matter of political geography, with the notable exception of elections, did not seem to have much to offer in the search for "spatial laws." The almost exclusive focus on the state was out of step with the urban and within-state concerns of the new human geography. Political science did not offer the inspiration for locational models that economics and sociology supplied to other subareas of the field. Textbooks continued to survey the geography of bits of politics, such as boundaries, capital cities, administrative areas, electoral districting, and geostrategies of the U.S., but with little interest in how the parts related to the whole. In failing to come to terms with the loss of its physical-determinist base, political geography had lost its way. At first blush, neither geography nor other fields had much to offer by way of salvation. Until the late 1960s the outlook was bleak. Yet help was soon on its way.

In the first place, we want to stress the role of external social and political conditions more than the intellectual efforts of political geographers or other scholars. The events of the late 1960s in North America and Western Europe brought practical politics to the fore in the everyday lives of both ordinary people and academics. The race riots, civil rights marches, Vietnam War demonstrations, and student rebellions of the time had a deep impact on social science. In all fields they brought political questions to the center of concern in ways that they had never been before. In geography this tendency was expressed in three ways. One was by bringing issues of the distribution of power into the analysis of economic and social phenomena such as residential segregation by race and class in American cities and the global distribution of economic development. These were no longer seen as purely market driven or a matter of free choice but the outcome of systematic bias in political institutions, such as local governments and school districts, managing the distribution of public goods. From this point of view, all of geography became political geography, at least of a kind. From the mantra of "states are everything" that had long animated political geography, geography as a whole confronted the claim that "politics is everywhere."

Another expression of the political quality of the period was the politicization of geography through public analysis of the field itself: who runs it, for whom, with what ends in mind? In the 1960s higher education had expanded massively in the United States and elsewhere. An important consequence was the recruitment into student bodies and the ranks of university teachers of people from social backgrounds hitherto largely excluded from this world. They did not always accept the norms of personal conduct and political outlook that had evolved in academia during the postwar years. They also challenged the benign acceptance of statehood and international geopolitical hierarchy as "facts of nature." A significant example of this comes from France, where Yves Lacoste crafted a new "geopolitics" based on a critique of the old version as essentially geography's contribution to the war making of powerful states and insisted on the irreducible spatiality of politics within as well as between states. Lacoste's "géopolitique" is one of the most obvious fruits of the explicit politicization of political geography to emerge in the early 1970s (Claval 2000; Hepple 2000).

In addition, there was a dramatic increase in academic mobility, particularly within the English-speaking world, that brought people with very different social and national backgrounds to those countries, such as the United States, Canada, and Australia, that were most active in recruiting new graduate students and staff. The politicization was not simply disinterested or idealistic, however, because power within the field over staff appointments, publication outlets, professional reputations, and external influence was also very much at stake. The expansion of universities had the effect of dramatically increasing the number of students and faculty, increasing the volume of research and publication, and encouraging new intellectual trends and "niches" so as to find employment for graduate students and gain access to funds for research and graduate financial support.

Finally, political geography was discovered by a new generation without much background or interest in the old studies. They tended to view the subfield naively as something they were inventing afresh rather than an old enterprise into which they should be inducted. Without the old obsession with defining disciplinary boundaries, they searched around in other fields for inspiration and found it in economics, radical sociology, anthropology, economic history, and even political science. At the same time, they embarked on explicitly redefining the field as the geography of politics more than the geography of states. Whatever the geographical scale or context—urban, regional, national, world-regional, or global—as long as power pooled up in some places, political organization privileged some in some places over others elsewhere, and territorial boundaries were used to exclude and include, political geography had research questions of interest.

Second, however propitious the time, nothing would have happened without the contributions of the people making the new political geography. Three types of contribution were crucial. The first, associated particularly with Kevin Cox, Roger Kasperson, Ron Johnston, Peter Taylor, and Richard Morrill, involved the revitalization of electoral geography. The focus here was on quantitative detection of the local dependence of election results in terms of local social indicators indicating patterns of social interaction and the ways in which electoral systems and districting methods affect the overall balance between political parties after elections. Elections provided good information upon which to postulate and measure the impact on social life of local and regional differences. Rather than just ends in themselves, therefore, when examined geographically elections provide an entrée into understanding the social dynamics of politics without presupposing either the individualized voters or national homogeneity of much orthodox political science.

The second, also associated with Cox and emerging in the early 1970s, was a focus on urban conflicts. The main idea was that much urban politics is about who gets what with respect to public goods and public bads and that this is determined largely by where you live (Cox 1973; Cox and Reynolds 1974). This is because most public goods and services are delivered by municipal governments to local areas, and public bads, such as air pollution and noxious land uses, are located closer to people who cannot afford to live elsewhere and do not have the power to intervene effectively in "locational conflicts" to keep such public bads at bay. This was developed in the 1970s as a welfare approach to political geography moving out from the early urban emphasis to examination of the distribution of goods and bads at a range of geographical scales, using the same logic for each (Cox 1979).

Finally, the new human geography attracted the interest of scholars in other fields, such as political science, diplomatic history, and sociology, who began to contribute their own type of political geography to the academic literature. Stein Rokkan, a political sociologist; Immanuel Wallerstein, a historical sociologist; Alan Henrikson, a diplomatic historian; and political scientists such as Bruce Russett, Richard Merritt, and Jean Laponce combined their own disciplinary expertise with certain geographic concepts—such as center-periphery, region, and distance-decay relationships—and methods—such as grouping procedures to show UN voting blocs (Russett), schematic maps showing the territorial structure of statehood in world regions such as Europe (Rokkan), and series of world maps to show historic shifts in the perception of global centrality (Henrikson). Geographers such as Jean Gottmann, Paul Claval, and Owen Lattimore actively encouraged the interdisciplinary rapprochement through their participation in the Permanent Research Committee on Political Geography of the International Political Science Association (IPSA). This organization played a much more important part in the initial revival of political geography than did the International Geographical Union (IGU), which remained mired in the talking-shop posturing of the Cold War. The term *political geography* was proscribed by the IGU until 2001. Interestingly, it is the nongeographers who helped most to reestablish the interest in geopolitics and international affairs that had been eclipsed in the postwar years. The geographers, for the most part, spent their energies until the 1980s on expanding the scope of political geography in new directions rather than attending to such issues as the rise and fall of dominant states, the geographical structure of empires, and the geographical origins and development of the European state system. Less hidebound by the heavy hand of political geography's past, the nongeographers were more capable of crafting political geography's reengagement with both the spatiality of states and the geography of world politics.

By 1980 there had been a remarkable resurgence in political geography as a multidisciplinary enterprise after the long dormancy of the postwar period. This was to deepen in the course of the next decade both with respect to the amount of new research (manifested in the founding of new journals, particularly *Political Geography Quarterly*, later *Political Geography*, in 1982 by Peter Taylor) and the theoretical development of the field beyond spatial analysis in a number of directions, especially those associated with political-economic perspectives and, more recently, postmodern approaches of various types. It is to the three theoretical frameworks and the various subjects of inquiry in contemporary political geography that we shall now turn.

THREE "WAVES" OF THEORY

The 1960s was a time when geography underwent a "spatial turn." Space or distance was defined as the field's "variable" (like the "economic" in economics). Older definitions, such as the regional and the environmental were (temporarily) eclipsed. Not surprisingly, the initial reinvention of political geography, coinciding with this development, also took on a spatial-analytic cast. The search was on for distance-decay effects in the influence of voters on one another in the choice of political party and externality-field effects from noxious facilities and "undesirable" neighbors on decisions to participate in neighborhood political protests. There was also considerable interest in the impact of districting methods on election results, the modeling of distance-decay effects on the possibility of war breaking out between states, and the spatial organization of local and regional governments. These interests and the spatial-analytic approach they tend to share have persisted, and revived with the arrival of GIS, digital mapping, and spatial databases (e.g., O'Loughlin 2003; Agnew et al. 2008). Though today they are nowhere near as prevalent as they once were.

The spatial-analytic approach is exemplified by the theoretical logic provided by Kevin R. Cox and David R. Reynolds (1974). After a brief review of the neglect of geographical considerations in studies of power and conflict, the authors identify two factors that they see as leading to an increased concern for space in political studies: the increasing effect of externalities (effects on others who are not parties to a transaction) on people in industrial societies and the adoption of "systems" perspectives in political science that tend to increase the attention given to outcomes of the political process (who wins where) rather than just the political process itself. So even if politics in the past could be thought of as "spaceless," this is no longer the case. The focus on urban settings and the spatial patterns produced by externalities (think of pollution plumes from factory chimneys and the benefits from living inside the catchment boundary of a high-quality high school) firmly distinguished the approach from previous ones. The logic, however, is not specific to either the urban scale or to such local external effects. It can be extended to interpret national and international relations in similar terms, as can be seen in Cox (1979).

During the 1970s geography experienced something of a turning away from the dominance of the spatial-analytic perspective. In the context of an extended period of political and economic crisis in many Western countries, many turned toward theoretical perspectives that could encompass the current situation as well as offer fresh understandings of old topics. A revived political economy proved especially attractive. Drawing in particular from Marxist and neo-Marxist writing in political economy, scholars framed political-geographical phenomena in terms of global patterns of uneven development and the processes that they claimed produce them. One variety of this approach, that of the world-systems theory associated with Immanuel Wallerstein, proved particularly influential in political geography. It has been popularized in Peter J. Taylor's textbook (1989) and in numerous publications by Taylor and others. Theoretically eclectic, drawing its main tenets from such different thinkers as Fernand Braudel, Karl Marx, and Karl Polanyi, this perspective tends to explain most other phenomena with respect to where they are located within a global division of labor (core, periphery, semiperiphery) produced by the historical workings of the capitalist world economy. Under the political-economic rubric, however, can be found a range of perspectives, some adhering closely to an orthodox Marxism (e.g., Harvey 1993) privileging processes of capital accumulation pure and simple and others exploring the autonomous powers of states (e.g., Mann 1984; Skocpol 1994). What joins them together is their view of space as a surface upon which political-economic processes (whatever the specifics) are inscribed and embedded but which is nevertheless essential to the outcome of the processes (for example, providing the "spatial fix" to the declining rate of profit by moving investment from one place to another, defining the spatial limits to state autonomy, etc.).

The 1980s did not relieve the sense of crisis, but this time the consequences were even more profound for political geography. Over the years many thinkers had questioned the pretensions of "grand theories" and "master narratives" in the social sciences, pointing out how they "overreached" the empirical evidence used to support them. Others had suggested that knowledge was more a product of language, conventions of study, and the relative power of different scholars than of independent "facts" about the world. As a result, one critique, associated particularly with feminism and postcolonialism, came to emphasize the partiality or "situatedness" of knowing; knowledge is a function, at least in part, of the standpoint or subject-position at which a scholar is located, particularly the historical experience of power relative to others and, thus, the capability to tell your story and those of others like you (Krishna 1993). Another critique came to play up the role of language and writing in offering meaning to readers. From this point of view, the world is written, not discovered or explored (Barnes and Duncan 1992). In other words, in this poststructuralist or deconstructionist view, writers recycle metaphors and tropes rather than discover new knowledge. Finally, some identified the tenuousness of all claims to tell "stories" about other people and their places. Even "emancipatory" narratives, stories told to benefit the interests and identities of others, involve a quest for transcendence that disciplines and limits the aspirations of presumed beneficiaries. In this postmodernist view respect for irony, ambiguity, and the paradoxes of existence remain as the only guarantees against imposing order on others. To the extent that it is possible, one looks for the stories that groups share to understand their self-constructions. At the extreme, however, one can never "speak for others" (Alcoff 1991–1992).

These are distinctive critiques, however, and the use of the term *postmodern* to cover all of them is problematic (Duncan 1996). In this context, it is used to convey the sense common across all of them that knowledge is both political and deeply compromised by the language and social conventions of academic fields and historical-geographical contexts (Gregory

1989). Some writers move uncertainly across the critiques in their work—feminism versus deconstructionism, for example—without much sense that these are philosophically distinctive. In moving beyond critique, the focus increasingly has been on the question of "identity"—the relation of the self to larger social groups and the world at large—and how answers to it depend on cultural context (language, understandings, experience, etc.).

The three "waves" identified above continue to flourish within contemporary political geography. At the same time that they have "washed over" the field, they have also helped to stimulate and refine a number of distinct areas of study within political geography. To the long-established themes of geopolitics and the spatiality (or geographical organization) of states and other polities three other themes have been added since the 1960s: geographies of social and political movements (including electoral geography), places and the politics of identities, and geographies of nationalism and ethnic conflict. The new themes reflect the reinvention of political geography well beyond the confines of its early twentieth-century protagonists and their dual obsessions with geopolitics and state territories.

THE THREE THEORETICAL WAVES AND THE 2008 U.S. PRESIDENTIAL ELECTION

To help understand the main features of the three ways of thinking about theory, it may be useful to provide a single empirical example and identify the distinctive ways in which the three types of theory might address the phenomenon in question. The U.S. federal system has two interesting and vital features. One is that it is fiscally redistributive: the federal government collects revenues (primarily the federal income tax) from people and businesses across the country and then through its expenditures on roads, schools, welfare, and defense spending sends the money back to the states both directly and indirectly (to municipalities and specific projects). The other is that presidents of the United States are elected through the states in the form of the Electoral College rather than directly: there is a weighting intrinsic to the system that by adding every state's two senators (independent of state population size) to its number of U.S. Representatives departs from a strictly popular vote. Nevertheless, the popular vote usually (but not always, as we know from 2000) parallels to a considerable extent the so-called electoral vote. In 2008 Democrat Barack Obama defeated Republican John McCain in both popular and electoral votes for the presidency of the United States. Obama's level of support was uneven, ranging from a low of little over 30 percent of the vote cast in Wyoming to over 90 percent in the District of Columbia. Various ideas could be and have been put forward for both why Obama won and why he won so unevenly. One plausible hypothesis, particularly in light of how Obama has been portrayed by many of his political opponents since his election, is that he represents the interests and identities of those in favor of a relatively powerful federal government. Presumably, this would include those who benefit most on a per capita basis from the redistributive activities of the federal government. If this were so, one would expect to see a high geographical correlation between the overall vote for Obama in 2008 by state and the amount that each state receives back from the federal government relative to what it sends to Washington by way of revenues. Graphing the two variables and mapping their intersection by state, however, reveals a pattern other than what we expect (figure 4.2). Federal spending relative to revenue collection is typically higher in states that voted much more highly for McCain than for Obama. Texas is the *only* state that voted highly for McCain that receives back less than one dollar for every dollar sent to the federal government. In fact, the correlation between the two variables in question (Percent Obama Vote 2008 by State and Federal Spending per Dollar of Federal Taxes Received from States 2005) is a *negative* 0.44. Obama typically did worse in states that get more money from the federal government relative to what they send than he did in states that receive less back than they send. How can this seemingly counterintuitive relationship be explained?

Let us take each of the three types of theoretical perspective in turn to see how they could account for the map and graph in question. From a spatial-analytic perspective we would have to reject the hypothesis outright. From this viewpoint, negative results like this one are not very interesting. We need to immediately turn our attention in a more propitious direction toward alternative variables that might better account for Obama's national success based on uneven support across the country. Spatial analysis is precisely about finding overlaps in geographic patterns that are at the very least suggestive of some causal relationship between the variables that are mapped. Only if the dots were all or largely confined to the darker shaded states in figure 4.2 could we get excited. When there is no overlap, as we have here, then we have a problem. From this perspective, the result just is not interesting or worthy of further study. At best, we have a problem of an underspecified theoretical model: all sorts of unknown variables could be at work here pushing the results in the negative direction we see on the map. Since we have no data on these, there is nothing more we can say. We are entirely limited by the variables we have and the cross-mapping strategy we follow to make sense of how they relate.

From a political-economic perspective everything is much more complicated. Any and all graphed and mapped data are invariably limited and problematic in getting at what we are most interested in. In this case it is all about how the role of the federal government figured if at all into Obama's election. The data reported on the map is only the tip of the "processual

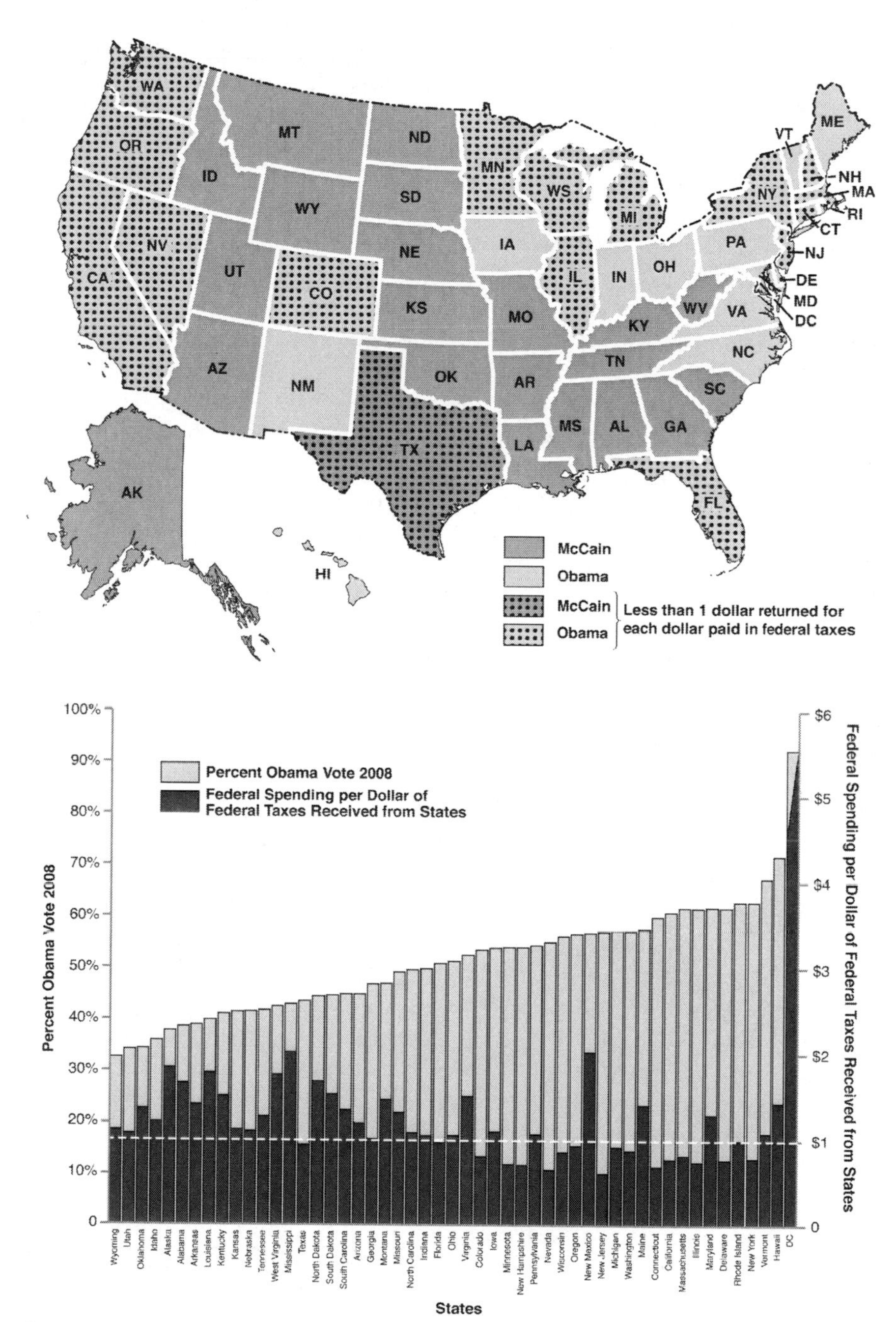

Figure 4.2. Percent of the 2008 presidential vote for Barack Obama and federal spending per dollar of revenues collected in 2005 by state.

Source: Authors

iceberg," so to speak, that we need to explore in order to explain what is going on. Yes, the states differ in terms of federal spending and voting for Obama, but these are related to one another historically and institutionally. They cannot be thought of simply as overlapping map patterns. Attention focuses on two factors: the characteristics of the states in terms of revenues/ expenditures and how the states are involved differentially in the allocation of expenditures through their relative power in the federal government. First of all, as reported in figure 4.2 there is no control for economic development differences between states. Irrespective of presidential voting inclinations, poorer states can be expected to receive more funds relative to population than richer ones. The results, therefore, are at least partly an artifact of this difference. Some states are getting more money because they are poorer, not because they have a higher opinion of the federal government than states that get less and ipso facto would prefer a president who is in favor of more federal spending. Second, as is well known, the return of funds to states has much to do with the ways in which the U.S. Congress is organized. States that have powerful House and Senate Committee Chairs, for example, historically do better by federal spending than those that do not. Understanding the workings of the national economy in producing uneven economic development and political institutions in lending further bias to the distribution of public resources would suggest that there is no direct relationship between federal disbursements and votes for the presidency. Indeed, it would be the effectiveness of congressional representatives more than who is president that matters most in terms of the role of the federal government relative to the states. From this perspective, figure 4.2 is a complete oversimplification of what is a much more complex process that can never be reduced to a two variable map of the U.S. presidential election. The U.S. presidential election simply does not matter that much in relation to what happens between Washington and the states of the United States.

Finally, under a broadly postmodern perspective the map and graph in figure 4.2 can be interpreted as trying to tell one story but as probably obscuring at least one more interesting alternative story that can be told, even though it is a "negative" one from the spatial-analytic perspective. In neither case is an absolute truth held as being at stake. Mapping and graphing may take effort, but it is not clear that they are anything more than ways of telling stories or constructing narratives that help us make sense of the world (and thus allow us to work in the world) but which are not more "truthful" than more anecdotal ones that inspire the actions of powerful figures who bring about change, notwithstanding the lower "scientific" status of their stories. It is the relationship of knowledge to power (why was information collected, whose story does it tell, etc.) not the nature of the knowledge itself (what sort of information is involved, how was the information analyzed, etc.) that thus attracts attention. The story that figure 4.2 is supposed to tell is about Obama as a federalist president with federal spending/expenditures standing in as a surrogate measure for ideological affiliation to federalism (the doctrine of a more powerful central government relative to the states enunciated by such founders of the United States as John Adams and Alexander Hamilton). First of all, as the map shows, Obama did much worse by and large in the states receiving the best deals from the federal government. These are in fact those parts of the country that have also been most hostile historically to an expansive role for the federal government: the states of the old Confederacy and the libertarian strongholds of the Intermountain West. The message from the map is loud and clear: you cannot read political ideology or identity from a map of presumed economic interests. Second, the federalist narrative is not the one that informed the election campaign of either Obama or of his opponent in 2008. It emerged after the election (as noted in the description of the hypothesis above) when Obama's opponents on the far right of the Republican Party in the so-called Tea Party Movement abandoned the big-government conservatism they had supported under President George W. Bush by portraying Obama as the singular source of the federal government's huge deficit (when at least half of it came from Bush-era wars and Medicare prescription benefit spending and much of the rest from the post-2008 effort at stimulating the U.S. economy after the financial crash of the period 2007–2009). Nothing much of that deficit can be put down to Obama's presumed federalist inclinations as such. Beyond its folkloric invocation of early American history as if it were ongoing today (Lepore 2010), the story of an America locked into two warring camps has other more recent sources. The narrative also involved a heroic effort to situate Obama ideologically within a certain discursive framework as the inheritor of civil-rights era federal expansion and as an African American within a "liberal" conception of the role of the federal government. In harking back to the "states' rights" rhetoric of such icons of southern segregationism as Alabama Governor George Wallace in the late 1960s and the race-baiting "southern strategy" for recruiting disaffected southern whites to the Republican Party in the years when Richard Nixon was president of the United States, this narrative reached its apogee in the congressional election of 2010 and in the debate over raising the national debt ceiling in 2011. Figure 4.2, therefore, can be mined interpretatively to tell us a considerable amount about the *lack* of connection between economic interests and the various divisions and factions in contemporary American politics, the uses of a certain weird reading of heroic periods in U.S. history to inform present-day debates (as in the Tea Party analogy to the period of U.S. independence from an overweening imperial power, Great Britain, and the language of federalists and antifederalists it engendered), and the use of after-the-fact causal arguments (such as projecting the Tea Party argument of 2010 back onto the 2008 election and classifying politicians as federalist and antifederalist, if often confusing the latter with supporters of the pre-1787 Articles of Confederation) to find an explanation for the election's outcome. The postmodern perspective, then, is both a critique of the other efforts at explanation (and how they fail to link with the actual ongoing politics of the time but stand apart from and above it to offer their explanations) and a way of providing *understanding* in terms of the narratives (dominant and subordinate) that animate actual politics, including those borrowed from the other theoretical perspectives.

PERSPECTIVES AND SUBJECT AREAS

Table 4.1 attempts a cross-tabulation of the three types of theoretical perspective with the five subject-areas of political geography. In each entry in the table is an example of an article or book that conforms to a particular combination of elements. In describing these in the following pages, the hope is that the structure of approaches and subject matter in political geography will become apparent. Some of the examples are described relatively briefly, while three, those of Rokkan (spatial-analytic), Osei-Kwame and Taylor (political-economic), and Ó Tuathail (postmodern), are given somewhat greater attention. These three represent key examples of each of the three types of approach to political geography. All of the examples used here appear in a reader (Agnew 1997b) that can be consulted to see them in their entirety or as substantial extracts of longer publications.

Geopolitics

Until the 1980s there was little by way of any revival of interest in geopolitics. The explicit geopolitical language used during the years of the Nixon Administration in the U.S. (1968–1974), however, did encourage a return to the topic (Hepple 1986). One of the best pioneer attempts at reengaging between U.S. foreign-policy making and geopolitics is a study written by Alan K. Henrikson in 1980. The author uses a center-periphery model of global

Table 4.1. Matrix of theoretical perspectives and subject areas.

	Perspectives		
Subject Areas	*Spatial-Analytic*	*Political-Economic*	*Postmodern*
1. Geopolitics	Henrikson	Corbridge	**Ó Tuathail**
2. Spatiality of States	**Rokkan**	Mann	Krishna
3. Movements	Bennett and Earle	**Osei-Kwame and Taylor**	Routledge
4. Places and Identities	Murphy	Wacquant	Forest
5. Nationalism	Conversi	Williams	Johnson

political centrality to argue for shifts in the perceived centrality of the U.S. to world affairs on the part of Americans. In other words, the history of U.S. foreign relations is the history of a shift from perceived peripherality to perceived centrality. Henrikson traces this history using a chronological series of maps in which the U.S. moves with fits and starts from the edge to the center of the world. After commenting on how ill at ease Americans can be in terms of their relative global status, and on how little they know about anywhere else, he shows how a policy of global engagement emerged slowly out of continental and hemispheric renderings of America's place in the world. He identifies the Spanish-American War and World War II as particularly important in shifting the U.S. to "world leadership" and a sense of geographical centrality. He also notes the failed attempt by President Woodrow Wilson after World War I to engage forcefully with the rest of the world, suggesting that Wilson had run far ahead of the American population in this respect. A map that represents the decisive movement of the U.S. to global centrality shows the U.S. going from hemisphere defense to global offense following the Japanese attack on Pearl Harbor in 1941 (figure 4.3).

Other writers, such as Saul Cohen (1973), John O'Loughlin (1986), Patrick O'Sullivan (1986), and Jan Nijman (1992), have also made important contributions to geopolitics using spatial-analytic perspectives, if with less emphasis on either the perception of national centrality or the use of maps and more on the "objective" global factors—resource access, contiguity of competing states, and others—conditioning geopolitical relations.

One criticism of this approach would be its neglect of the world-economic context in which the geographical framing of foreign policies takes place. America's rise to global prominence owes more than a little to its economic position. Yet even as it has achieved global political centrality, the U.S. has also changed the nature of the competition upon which global competition has long rested. It brought its home-based focus on economic expansion to bear beyond its boundaries (Agnew 1999). As a result, it seems as if economic prowess is now as important in its own right as its translation into military might and political influence. Interstate competition now often seems to be as much or more "geoeconomic" than "geopolitical." But the change in global conditions may be more profound than this. Stuart Corbridge (1994), for example, claims that the global context has changed to the extent that states are no longer the singular actors of world politics. All manner of international regulatory organizations and private businesses are now important actors in world politics in their own right, while the past decade has also seen the economic rise of China, India, Brazil, South Africa, and Russia, making for even greater variety and complexity to the contemporary world "geopolitical economy" and challenging the dominance of its three global centers—the United States, Japan, and the European Union. As yet, however, the world economy is still dominated and linked by flows of capital

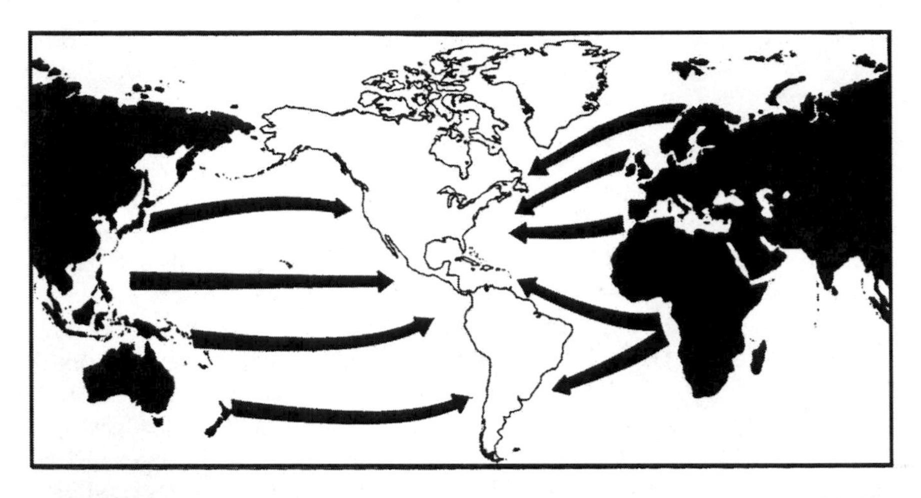

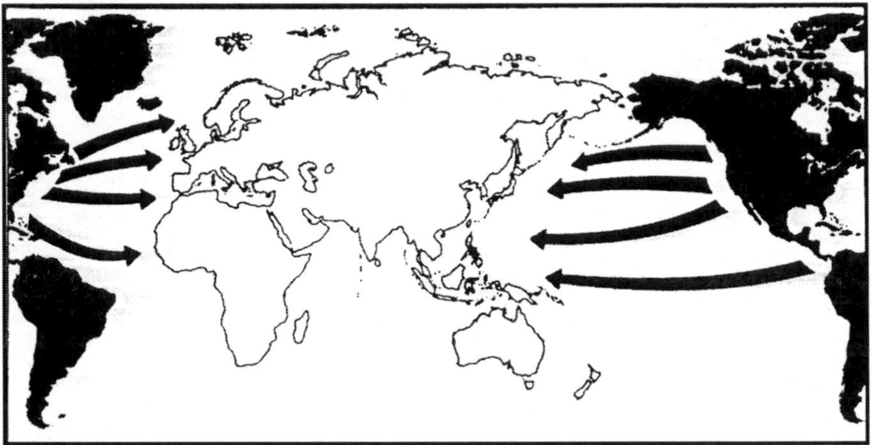

Figure 4.3. The future of the western hemisphere according to Nicholas Spykman (1944): from western hemisphere defense to global offense. Used in Henrikson (1980).

and by financial and trade connections that emanate from such centers as London, New York, and Tokyo to distant sites around the world. In this construction, the world is not driven by zero-sum (winner-take-all) competition between states but by relative abilities of governments at local and regional as well as national levels to insinuate localities and regions into the circuits of global capital (Florida 2011; Gilbert 2011). Many states in fact have little if any capacity to encourage or retard economic development. With such "quasi-states" and uncertain sovereignties among the more powerful ones, there are opportunities both for a less militarized world and for new political instabilities (such as those represented by international terrorist networks and religious-based political movements).

Therefore, this sort of historical political economy, arguing for the role of changing historical conditions on world politics, is a relatively recent innovation in political geography. More deeply entrenched are viewpoints that emphasize the permanent importance of the "geopolitics of capitalism," looking at evidence for contemporary globalization as only the latest in a series of reorganizations of business ("capital") at different geographical scales to resolve the long-term tendency of the rate of profit to fall without a constant search for a "spatial fix" (Harvey 1982). Meanwhile, other viewpoints, such as those proposed by world-systems theorists and advocates of cyclical models of history, stressed the emergence of new hegemonic states out of the ruins of past epochs because of, respectively, the clustering of technological advantages (Taylor 1989; Wallerstein 1993) or fewer military obligations relative to national economic capacity (Kennedy 1987). The 2000s U.S. War on Terror perhaps exemplifies this trend with major resources (including funds borrowed abroad) poured into military spending, thereby massively increasing the public deficit (together with the effect of the 2008 financial crisis bailouts) in the absence of raising revenues to cover the increased spending. As a result, the U.S. national debt has catapulted to three times its 2000 figure, while the percentage of it in the hands of foreign holders has increased to almost a third of the total. As the prospect of paying off this debt hinders overall growth of the economy, U.S. hegemony as it has been exercised faces a significant erosion, while a major shift in global power toward creditor countries, such as China, is already under way, signaling the advent of a global geopolitics less centered on a single dominant economic power (Agnew 2009).

As Corbridge (1994) notes early in his article, the end of the Cold War coincided not only with an upsurge in writing about geopolitics from a variety of perspectives but also with an explicit questioning of the geographical assumptions and language upon which the practices of foreign policy are based. A self-consciously "critical geopolitics" has developed in which the ways in which politicians and the media represent places and their strategic significance take center-stage (Ó Tuathail and Agnew 1992; Ó Tuathail 1996; Muller 2008). Informed by such poststructuralist writers as Foucault, Derrida, Virilio, Baudrillard, and Deleuze, emphasis is placed on deconstructing the discursive strategies used to make foreign-policy "situations," "crises," and "wars" intelligible with respect to both a global "big picture" and to past events seen as analogies to present ones (such as Munich/appeasement and Vietnam/quagmire in U.S. foreign-policy discourse). The primary actors surveyed are "intellectuals of statecraft," particularly political leaders and their advisors/seers, such as those from influential "think tanks" (McGann 2007, 2011). Some writers have tended to stress the ways in which geopolitical representations serve to anchor national identities as dominant narratives are popularized (Campbell 1992; Sharp 2000; Mamadouh 2008). Others have focused more on how such representations frame the conflicts that flare up to challenge the dominant actors in world politics. It is to one of these that we shall now turn, in a case study situated right at the end of the Cold War.

Gearóid Ó Tuathail (1993) provides a detailed analysis of the First Gulf War of 1990–1991 in terms of two main themes: (1) the placing of the Gulf "situation" in the big picture of the end of the Cold War and the desire of American political leaders to "re-territorialize" the U.S. as the sole superpower, and (2) the war as an example of the "dematerialization" of place through the use of new high-tech weapons and the immediate visibility of the war on television screens around the world as a type of "war game" in which the place the war was happening was incidental to the demonstration of high-tech capability.

The two themes are developed in relation to the speeches and actions of American politicians, particularly President George H. W. Bush. Ó Tuathail (1993, 6) acknowledges that the Iraqi invasion of Kuwait in August 1990 "had complex regional origins" but that the "subsequent narration of this event as the globally significant and threatening 'Gulf crisis' must be traced back to dilemmas created for the United States and Atlanticist security structure by the end of the Cold War." The Cold War had given a structure to both world politics and American society, from dividing up the world, targeting military forces, and organizing military production to containing political dissent. The Gulf crisis offered an opportunity to write a "general threat narrative" to replace the one that had been lost and thus to make the "United States" meaningful again as an enterprise in world politics. A series of "recurring inscription strategies" were used to this end by American politicians and official commentators. Those of particular importance in offering "reasons" for the U.S. intervention are twofold: (1) oil in general—the defense of Saudi Arabia, oil reserves, and supplies—and the "chokehold" that Iraq would have on the world oil supply if allowed to occupy Kuwait, and (2) the pursuit of a "new world order" based on the rule of law and the defense of existing sovereignties (such as Kuwait's). Those of importance in situating this crisis and war in relation to past American experience included analogies with World War II (in particular, with Iraqi dictator Saddam Hussein as Hitler) and Vietnam (with the "syndrome" of popular American opposition to colonial-type wars finally overcome by the prospect of victory in the Gulf).

This discursive emphasis is supplemented with an account of the role of television and military technologies in making the Gulf War a spectacle in the way other wars had not been. Drawing from the French war theorist Paul Virilio, who sounds weirdly reminiscent of early twentieth-century Italian Futurists, the Gulf War is seen as offering "provocative evidence of the eclipse of place by pace" (Ó Tuathail 1993, 18). Rather than containment and strangulation, the American war strategy relied on using aerial assaults by missiles and other munitions to overcome the defensive advantage of the Iraqis. This approach was visible to observers in the dramatic news footage capturing the pinpoint bombing achieved by the American-led forces. Ó Tuathail cautions against accepting the idea that this is a totally new kind of war in which all of the old rules no longer apply. But he does claim that the net effect of both the discursive and technological strategies pursued during the war has been to "efface place" in the sense, first, of dehumanizing and distancing from observers the truly horrific consequences of high-tech warfare and, second, of understating the rich cultural variety of the places subjected to acts of war. Whether this is all that new, of course, is in fact open to question.

Ó Tuathail reminds us that despite recent changes in the workings of the world economy geopolitics remains closely attached to the conduct of wars. As guide and inspiration, the modern geopolitical imagination reduces places to points on a map that then cease to have inherent qualities and only count in a calculus driven by global desire. This vision is not new but has a long genealogy going back to the European encounter with the rest of the world from the sixteenth century onward (Agnew 1998). The content of this vision does change, however. The challenge to the student of geopolitics is to detect what is changing at any particular moment and what is not and whether these can be understood in purely representational terms or reflect interests and identities masked more than revealed by discursive constructions. Language is as much rhetorical and persuasive as it is representational. It is thus not always as revealing of motivations and desires as might seem at first glance. The theoretical richness of Ó Tuathail's attempt at connecting discursive with technological power is revealing of how limiting reliance on one to the exclusion of the other can be.

Spatiality of States

Along with geopolitics this is the most established area of study in political geography, if usually under a number of different labels such as territoriality, geography of state formation, and the geography of administrative areas (including federalism). The spatiality of states refers to both the external bounding and the internal territorial organization of states. Studying boundary claims, frontier disputes, and territorial organization have long been important research areas in political geography (Kasperson and Minghi 1969; Sack 1986; Newman and Paasi 1998). There is also a large body of writing on local governments as substate actors and the specific geographical features of federal systems (Paddison 1983). This kind of work continues to flourish. But increased attention is now being given to the spatiality inherent in the modern state system of competitive state sovereignties claiming total jurisdiction over their populations in their territories and how this came about.

This emphasis is not simply an intellectual happening independent of the times. The modern territorial state is now in question in ways that would have been unthinkable some twenty years ago. As the Cold War was unwinding and globalization was under way from the 1970s onward, a whole set of phenomena began to show the slow erosion of state sovereignty as it had been conventionally regarded: the onset of globalization in finance and production, the explosion of migrant and refugee populations escaping the collapse of states (from Afghanistan and Iraq to Sudan and Libya), the collapse of the "strong" states of the Soviet Union and Eastern Europe, the rise of supraregional (as in the European Union) and global (as in the IMF, UN agencies, the G8, G2, and G20) forms of governance to address contemporary problems highlighting a growing need for a more capable global governance, and the rapid increase in ethnic and regionalist conflicts within and between states (see chapter 5). The image of a "fixed" territoriality to political organization can no longer be taken for granted. This has encouraged a search for the historical roots of the territorial state as a form of governance and speculation about the limitation or displacement of state sovereignty as a governing principle of international relations.

The increasing popularity of less state-centered and coercive conceptions of power has also contributed to questioning established conventions about states and their grip on people and territories. These reflect changes in territorial organization and in historical understandings of statehood. For one thing, the expansion of regional tiers of government within states has led to a decentralization of power from previous concentrations in central bureaucracies in capital cities. For another, the sense that the history of statehood has been more complicated than dominant accounts in the social sciences have made it seem has inspired perspectives more sensitive to the limits than to the strengths of state power. The world contains a variety of state forms that cannot be reduced to a single model of statehood. States have never had an exclusive monopoly on the means of violence that much political theory might lead one to expect.

The idea that there is a necessary link between political community and territory is an old one in Western political theory. But only with the rise of the modern territorial state in sixteenth-century Europe was there finally a close affiliation between the two. Only since then have citizenship and territory been conjoined. This connection has become so taken for granted that much debate in political geography has assumed that territorial sovereignty is a realized ideal and turned to questions about the character of the state apparatus or political institutions associated with different kinds of state (capitalist or socialist, democratic or authoritarian, etc.). There is much to commend this approach, not least because questions about citizenship rights, access to institutions, and the role of states in legitimizing social divisions receive critical attention. This internal orientation, however, neglects the geographical underpinnings of statehood itself and the critical role of the internal/external distinction that is a vital attribute of modern statehood. The clear bounding of territories by states is one of the main differences between modern European-style political organization and the types of polity that prevailed in nomadic, clan, imperial, absolutist, and feudal societies around the world in the past.

Stein Rokkan (1980) spent many years investigating the historical roots of statehood in Europe. In his 1980 article he traces the process from the collapse of the Roman Empire down to the twentieth century. His purpose is to show the various spatial elements involved in state formation, particularly how core areas of states emerged, and the role of urbanization and relative location in Europe (central versus seaward and landward fringes) for state development. The experience of specific states is thus placed within the experience of Europe as a whole. Not for Rokkan was the dominant strategy in previous studies of taking an ideal case—typically, France or England—and seeing all other states as imitators of the prototypes.

The main characteristic of Rokkan's approach is that it is typological. He attempts to draw up a geographical template or simplification for Europe as a whole with north/south and east/west bands designating the regional and local impact of Europe-wide historical events such as the collapse of the Roman Empire and subsequent ethnic patterning of Europe, the spread of feudalism and the emergence of a central urban belt from Flanders in the north to Italy in the south, the effects of the Protestant Reformation and the subsequent Catholic Counter-Reformation, and the imposition of unitary and federal models of government. Consequently, the political map of Europe is seen as the outcome of a series of determinative Europe-wide processes—economic, military-administrative, and cultural—but with distinctive local and regional effects on the types of states and their internal structures.

Though there is no single "master map" on which all of the varied processes can be displayed, Rokkan does provide a spatial-analytic matrix for Europe in which the processes can be inferred from the outcomes as shown on the map (see figure 4.4). This is a classic spatial-analytic maneuver: reasoning back from present-day spatial form (what the map looks like) to the mix of processes that produced it. What makes it different from other research in this genre is its historical emphasis, seeing the map as the cumulative outcome of centuries of continent-wide "shocks" producing different effects in different places because of different prior accumulations of shocks. What is most notable on this mapping of European state formation is how it combines "system characteristics" (physical size, city structure, one dominant city versus many cities, linguistic unity/division, nation-state development, seaward-landward location, and city-state history) with actual empirical examples of combinations of

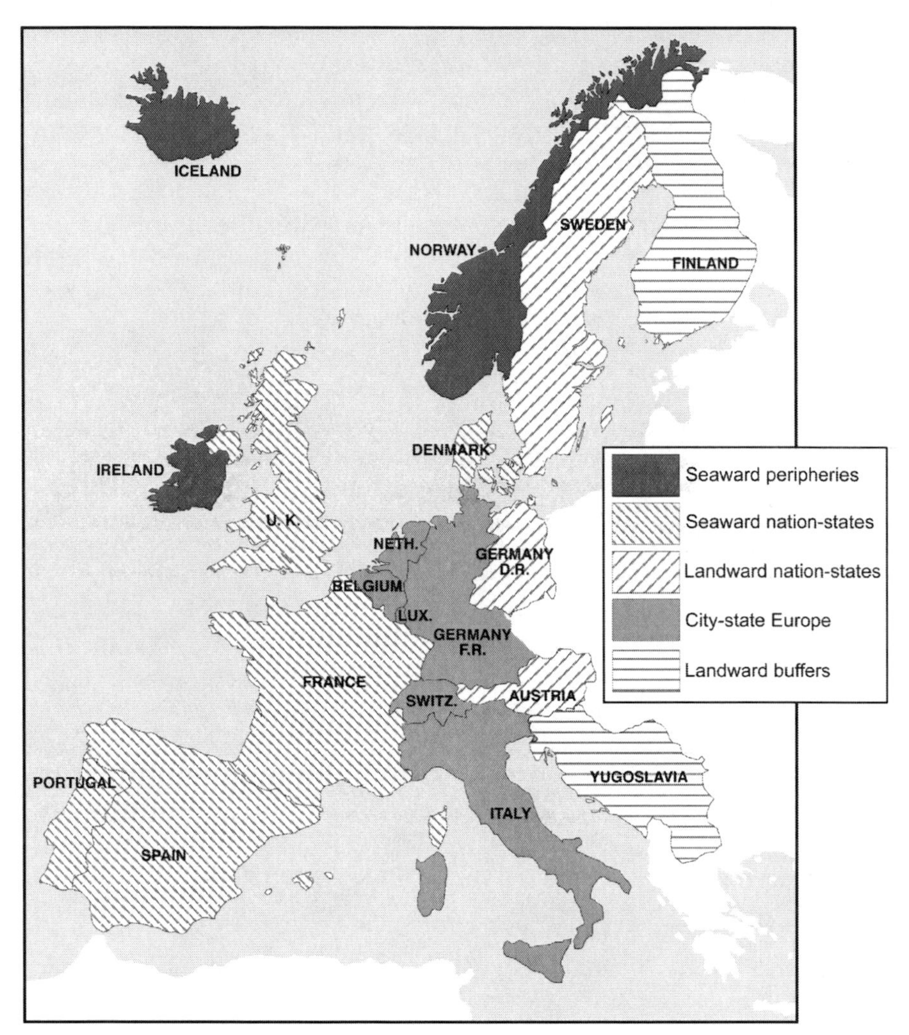

Figure 4.4. A map of Stein Rokkan's typology of political systems of twentieth-century Western Europe.

Source: Redrawn from Rokkan (1980, table 4.2)

the system characteristics. It therefore provides a means for both looking at Europe as a whole and for understanding the trajectories of its various parts, but always in relation to the whole.

In fact, research by Tilly (1990) and others (Tilly and Blockmans 1994) suggested that Rokkan's spatial-analytic approach missed important causes of the rise of territorial states in Europe. Their political-economic approach draws attention to relative military capacities (based largely on tax systems) and the differential rise of merchant and industrial capitalism in various parts of Europe. But they see states as emergent actors in their own right rather than simply the "instruments" of social classes such as merchants or industrial capitalists. If some, such as Wallerstein (1974) and Anderson (1974), have given more importance to the growth of trade and new social classes, though in different ways, others have emphasized the rise of state autonomy as one of the features of modern state development. Perhaps the best example of this perspective is provided by Michael Mann (1984).

Mann's (1984, 187) central point is that "the state is merely and essentially an arena, a *place*, and yet *this* is the very source of its autonomy" (his emphasis). What he means by this is that the modern state's very territoriality is what gives states (as opposed to groups from within society such as social classes) a high degree of autonomy. It does so because through the control over territory the state deploys infrastructural power. This is the result of the state responding to demands from within society and the society responding to the provision of infrastructural goods and services by the state. Only the state can exercise authoritative power within a circumscribed territory. Other social groups are lacking in this capacity. In addition, therefore, to the despotic power that groups can exercise if they gain control over states, infrastructural power accrues to the state itself, but only because it is a territorial enterprise. As a result, "Where states are strong, societies are relatively territorialized and centralized" (Mann 1984, 213).

Territoriality is a necessity for complex agrarian and industrial societies, therefore, because the state penetrates into the life of social groups as these groups give up powers to the state in return for various favors. In turn it is the very territoriality of the state that guarantees the state a degree of autonomy in relation to society. Its powers cannot be reduced to those of any one group. The state's capacity to do things for groups that they cannot do for themselves (infrastructural power) is the reason for this. This is distinctive from the despotic power exercised by state elites, which derives from their social roles, is usually precarious, and cannot insinuate itself like infrastructural power into the routines of everyday social life.

Of course, the success of the state as an autonomous actor depends on the extent to which its population vests its powers with legitimacy. From a broadly postmodern perspective, an argument such as Mann's serves such a purpose (even if inadvertently). It naturalizes and normalizes "the state" as an actor in people's lives. A postcolonial perspective would add to this the violence that statehood brings in its train. The experience of political independence for the former colonies of European empires serves to bring attention to the arbitrary nature of the bounding process involved in statehood and the difficulties boundary marking entails for those living at the borders. People must choose one side or the other; there is no recognized borderland identity, only competing state ones. Sankaran Krishna (1994) uses the metaphor of "cartographic anxiety" to convey how discourses about an Indian "nation" are used to define the borders of the Indian state. The "body politic" of India is defined in terms of a physical map that tries to conjure up a "historical original," a "homeland" that never existed prior to British colonial rule. So not only those at the borders of India are caught up in an exercise in spatial self-definition that is the essence of nation-statehood: abstracting from history a set of stable, legitimate boundaries that fix the history of the state in place and guarantee it a place in the future. The map, therefore, is an attempt at answering definitively the anxiety that comes from being "suspended forever in the space between the 'former colony' and the 'not-yet-nation'" (Krishna 1994, 508).

This returns us to the dilemma created by the fact that Western thinking about governance in general and democracy in particular is usually centered on the state (Legg 2011; Larner 2011). But state spatiality is based fundamentally on the exclusion of concerns about what is external as well as on the inevitable penetration of the state into society. As Machiavelli taught in *The Prince*, politics is possible only within state boundaries; reason of state operates beyond them. The spatial attributes of modern statehood, therefore, have more than passing relevance to the great questions of political theory about citizenship and democracy. They are at the heart of debates about the possibilities of democratic governance in a world in which many decisions affecting us all on a daily basis now increasingly emanate from distant seats of power beyond the reach of territory-based authorities (see chapter 5).

Geographies of Social and Political Movements

Much of politics is about mobilizing groups of people to obtain either public goods (and remove public bads) or redress grievances from political and economic organizations, of which the most important is usually the state. Public goods are policies and the provision of regulations and resources that benefit specific groups and places. Social movements often arise spontaneously around particular issues at specific locations. Sometimes they expand to encompass like-minded and similarly organized people elsewhere. They die out when a cycle of activity has passed or they become formal interest groups or political parties. Historically, and in the process of state formation, the "repertoires" of collective-action strategies used by movements tend to shift from the localized and sporadic to the national and systematic, from burning the hay ricks of landlords to mass demonstrations in capital cities, nationwide strikes, and boycotts (Tilly 1986).

But even when the issues they promote and the strategies they use are nonlocal, movements must still have roots somewhere. They cannot successfully mobilize unless they can attract recruits across a range of places. Of course, if local memberships opt for different goals and strategies, then mass mobilization becomes problematic. This is more likely when states are less centralized and local autonomy provides alternative institutional outlets to a focus on the center (Tarrow 1994, 62). The shared experiences and social interaction of living together in places provide a major stimulus to join movements. Absent such social incentives provided by living together in places, people are susceptible to leaving political action up to others. This is the so-called free-rider problem: that you can benefit from the exertions of others, and so have no incentive to participate yourself. That so many people do engage in political action of various types is testimony to the role that the regular rounds of everyday life exert in socializing people into politics (e.g., Garmany 2008; Nicholls 2009).

The period since the 1960s, particularly during the years 1965–1984, has been one of intense social movement activity around the world. Like previous periods of intense activity, such as between 1880 and 1910, many people have chosen to act with others in pursuit of common goals that might in other historical conditions have appeared unrealizable. Increasingly, such goals are not national in scope, just as national organizations and institutions remain the main targets of political action. From environmental problems to human-rights questions, new social movements have grown to address issues that do not admit of ready national-level solutions. Yet the difficulties of organizing transnationally are such that states remain the major "opportunity structures" within which social movements operate.

Many political parties, particularly progressive ones and ones on the extreme right, have begun life as social movements. So there is a continuum of movements running from localized, ephemeral social movements at one end to formal, institutionalized, and state-oriented political movements at the other. Political parties have received much more attention from political geographers than have the less formal social movements. This has largely methodological roots. Parties run in elections, so election results can be used to make inferences about the nature of support (which social groups support which parties) and the incidence of political ideologies among entire voting populations. There is now a large body of research in electoral geography showing the geographical covariation between political parties, political ideologies, social groups, and specific places. An important thrust of this research has been to identify the ways in which places mediate between political choice and social groups. Distinctive geographies of political parties, therefore, are the result not simply of a coincidence between where certain social groups reside and votes for different parties but of the way places structure political ideologies and affiliations (Agnew 1987, Miller 2000).

The structuring role of place in politics is the focus of an article by Sari Bennett and Carville Earle (1983). It offers a geographical analysis of an important issue in American political history: the failure of a socialist party to take root at the turn of the twentieth century, the last great period of social movement activity in the United States before the present post-1960s one. Previous interpretations of this failure have isolated the role of ethnic divisions among American workers, the general prosperity of American workers, or the faulty political tactics of socialist leaders (particularly their strident rhetoric and maximalist demands for total social transformation). Bennett and Earle prefer to emphasize the geography of the Socialist Party vote across the northeastern United States in the presidential and congressional elections of 1912, tracing the basis of success to the sedimentation of trade union or "labor power" in the years after the Civil War. In a statistical analysis amply illustrated by maps, they reason backward from the spatial pattern of support for the Socialist Party to what might plausibly have brought about the failure of the party either to expand nationally or to build on its initial places of strength. They identify two factors operating differentially across the northeastern United States that undermined the prospects of the Socialist Party in presidential and congressional elections: the increasing gap between skilled and unskilled wages in the larger cities, which divided the working class, and the industrial diversity of large cities, which reduced the relative numbers of unskilled workers in the heavy industries that stimulated political militancy. The Socialist Party had been successful largely in smaller cities with heavy industries. A base in the growing, larger cities eluded it and prevented its expansion onto the national scene.

This spatial-analytic reasoning is fairly typical of much research in electoral geography (e.g., Cox 1972; Johnston 1990; O'Loughlin 2003; Gimpel et al. 2008). The major criticism that can be directed at it is that it is reductive, searching for potential causes that operate differentially across space. In contrast, political-economic approaches frame spatial variation in political party or social movement activity in terms of an overarching theory of political economy. Party success or failure is seen in terms of the cycles of the economy and the balance of social forces at any point in time. A good example of this structural political-economic reasoning is provided in an article by Peter Osei-Kwame and Peter J. Taylor (1984).

The authors use a world-systems framework to argue that political parties in Ghana in the period 1954–1979 were constrained by the country's location within the global division of labor to compete over which economic strategy best served the country (loosely, import substitution versus basic commodity-export orientation) while appealing to different ethnic clienteles with divergent interests in commodity-export production (i.e., production of cocoa beans for export). Osei-Kwame and Taylor use a quantitative analysis of election results to identify a number of different subperiods in which different parties prevailed largely through isolating in opposition the party most representative of the Akan (Ashanti) ethnic group, closely identified with cocoa growing and, hence, most opposed to the "semi-periphery" economic strategy of import substitution pursued by the governing parties. The great political conflict throughout much of the period between Nkrumah of the Convention People's Party and Busia of the United Party was based around this division, with the former representing government centralization and protectionism and the latter decentralization, liberalization, and open trade. But it is also a politics of faction in that the centralizers must "shop around" to put together a coalition of places against the liberalizers whose base remains relatively stable over time. Osei-Kwame and Taylor interpret this as a pattern likely to be found in other "peripheral" states and former colonies in which political parties must work with a limited number of economic policy alternatives in the presence of major ethnic divisions.

Using a common set of electoral districts to facilitate comparison of elections over time, the authors use the technique of factor analysis to show that, with the exception of the 1969 election when the Akan (including Ashanti) cocoa regions emerged victorious, the election results favored parties stronger in the other non-cocoa-growing regions, with relative shifts among them over time. With military coups and governments replacing elected governments in 1966 and 1972, four electoral patterns are identified for 1954–1955, 1960, 1969, and 1978–1979 when elections did occur (figure 4.5). There is no "normal" or stable voting pattern in Ghana but a series of distinct mobilizations based on different coalitions across places in the country. In 1954–1955 a "national mobilization" for Nkrumah's party (positive factor scores) is scattered around the country, if with a clustering in his home area in the southwest, with opponents (negative factor scores) in the central Ashanti region (west-central) and in the north. In 1960 Nkrumah won a referendum on a new republican constitution and election as president. This produced a pattern of "state mobilization" similar to the earlier one but with opposition in the north, in Kumasi (Ashanti), and in urban areas in the south. In 1969 an Akan-versus-Ewe ethnic pattern emerged, with Nkrumah in exile and the Ewe party (from the southeast) representing the centralizers and the Akan the liberalizers. This time, and this time only, the liberalizers won, only to be overthrown in a 1972 military coup. Elections in 1978 and 1979 showed that a "new centralist mobilization" had emerged, pitting the Akan region (negative factor scores) against centralizing factions now dominant in the southwest and the north (positive factor scores).

The possibility of multicausal mobilization or spontaneous agency is excluded from a structural account of party politics such as this. The structural position of Ghana within the world economy and the correlation of production for the world market with ethnicity are seen as the driving forces behind political mobilization. Yet other authors have noted that resource conflicts, governmental interventions, and resistance to established power

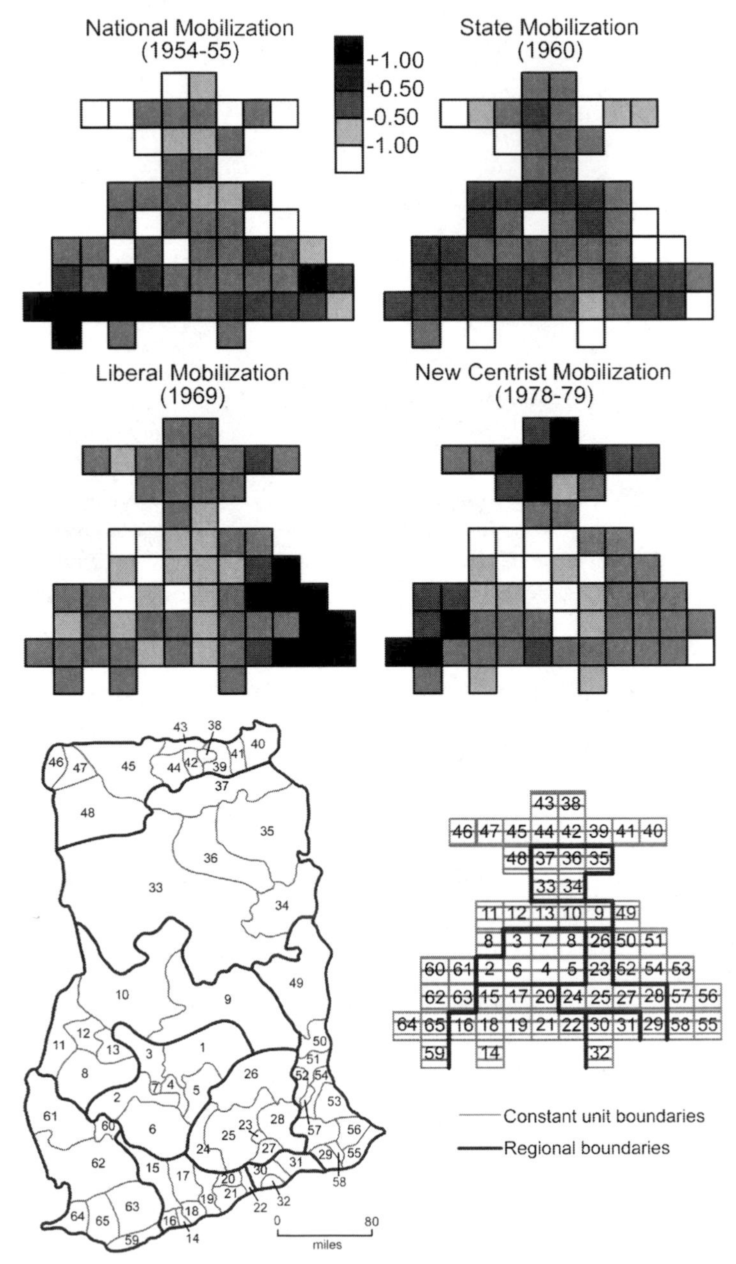

Figure 4.5. Osei-Kwame and Taylor's map of constant units of elections in Ghana, 1954–1979 and cartograms showing the changing "factor score" patterns for a T-mode factor analysis. Constant units are constructed by combining electoral districts to produce a common area database for 1954–1979. The cartogram locates each constant unit as close as possible to its real geographical location. Each district is reduced to the same size on the cartogram. The factor score cartograms use the constant areas to map the clustering of votes for different political parties/groupings across the elections. The titles of the cartograms each indicate the salient characteristic of each election or set of elections.

Source: Redrawn from Osei-Kwame and Taylor (1984, 211, 213)

structures can arise even when political resources and opportunity structures are not richly available (Pred 1990; Staeheli 1994). From this point of view, the relative success of political parties and social movements cannot be reduced to location within the global division of labor.

Authors operating with a "postcolonial" strain of argument have been particularly prominent in questioning the singular importance of economic considerations in political mobilization. This sort of approach is represented in an article by Paul Routledge (1992), who provides a geographical account of one local social movement in India. He sees this movement as emanating from a "terrain of resistance" (a whole set of local conditions) against the coercive, cooptive, and seductive powers of the state. The movement in question has devoted itself to opposing the Indian government's attempts at establishing a number of military facilities (in particular the National Testing Range for missiles) in the Balasore (Baliapal) District of Orissa, northeast India. Routledge traces both the activities of the movement in taking direct action against moves to set up the testing range and the sources of local activism. He sees the movement as a response to the perceived "militarization" of Orissa by local groups most affected by this trend (figure 4.6).

His argument, however, is that there is a geography to the origins of this social movement. He identifies a set of local and locational conditions that have contributed to the mobilization of local people against state plans. Key among these is a local "sense of place" that cannot be reduced to either a set of local material interests or local externality fields, to adopt spatial-analytic terminology. The idiom of the movement is strongly connected to this feeling for or identity with place through the songs and dramas used in political activities such as demonstrations and setting up barricades. The case strongly suggests that collective political action, at least in India, involves cultural codes that are very place specific. Routledge gives a similar interpretation to other social movements in India, linking their growth to the democratization of Indian society and popular opposition to the depredations of an overbearing state.

The claim of Alexis de Tocqueville in *Democracy in America* that there are greater political opportunities when society is stronger than the state and when there is considerable "local patriotism" seems strongly supported by this Indian case study. Perhaps it is the withering of local opportunities and the decline in local identities that accounts for the recent erosion of political participation in many Western countries. At the same time, the geography of political organization (the ways in which structures of governance, political parties, and social movements are organized over space) seems increasingly out of kilter with the cultural complexity and shifting economic geography of the time. Whether connecting in cyberspace or opening up institutional channels at local and supranational levels will help to revive

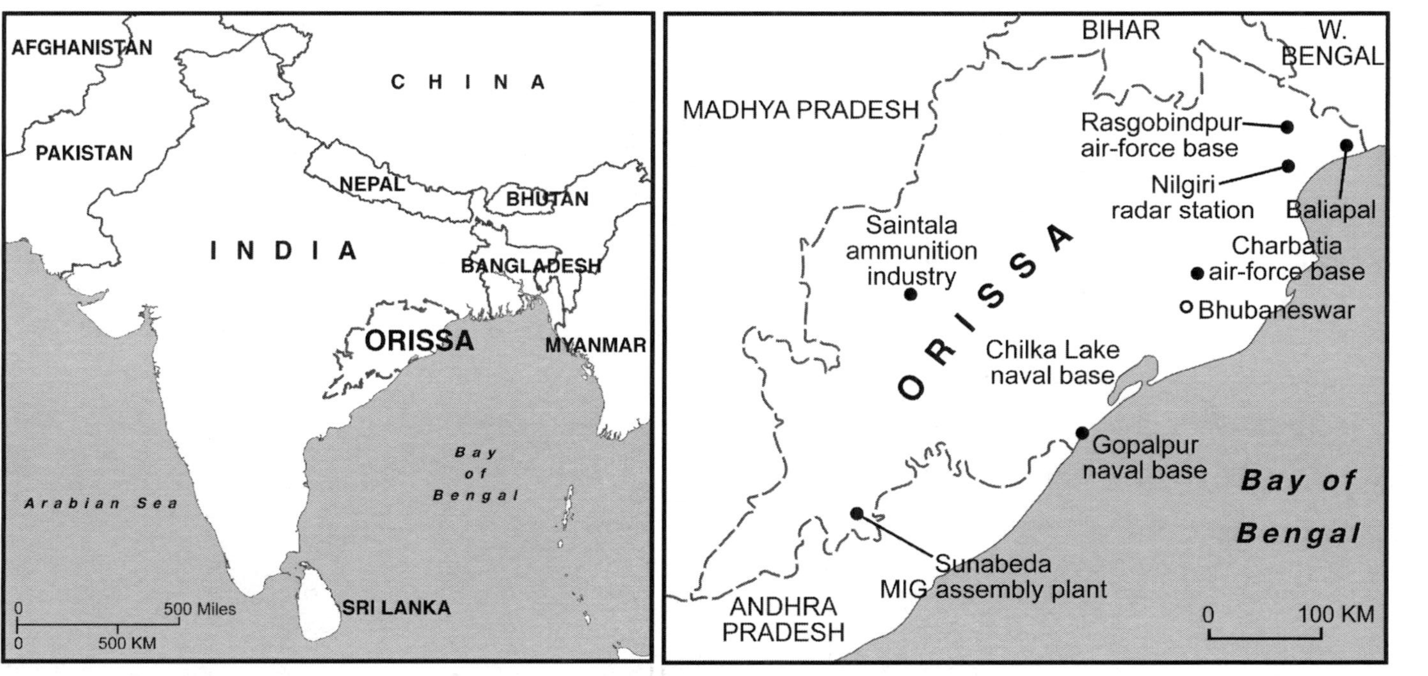

Figure 4.6. **Location of Orissa in India and the location of Baliapal and military sites in Orissa.**
Source: Redrawn from Routledge (1992, 226)

political participation remains to be seen. Of course, the decline in political participation may also be a cyclical downturn more than a structural trend. We have certainly seen that before. The trend of growth in social movements from the 1960s until 1984, particularly marked in Europe and the United States, may have been unsustainable.

Places and the Politics of Identities

Many of the social movements for racial civil rights, ecology, women's rights, gay rights, and religious purposes that sprang up in Europe and North America beginning in the 1960s are not simply about satisfying "interests" but also, and more importantly, about "identities." They are about the struggle for public recognition of identities that have been either stigmatized or suppressed within the wider society. Much contemporary politics, therefore, is a politics of identity, involving struggles over the recognition and legitimacy of different social identities. The seemingly instrumental character of older social movements, such as labor unions, has obscured the degree to which they also involve struggles for recognition and self-respect. Nevertheless, the self-conscious concern for identity is distinctively modern if not entirely recent (Calhoun 1994). When all-encompassing "identity schemes," such as kinship, prevail within a society, there is little problem. Only when we live in a world in which social-relational networks are diffuse and there is limited cultural consensus do individual persons face the difficulty of identifying who they are in relative isolation. Identity politics is about struggling to establish the recognition of collective differences in identity within a society where those differences are either not sanctioned or unacknowledged.

Identities are created through the stories people tell themselves and others about their experiences. In this way people come to see themselves as part of larger collectivities with common histories. People present themselves to one another in terms of stories and tell stories about one another. These narratives are attempts at creating a unified self, one that makes the self intelligible. Lives and stories are intertwined to become identities. From this viewpoint, identity is about the connection drawn between a "self" and a community of communicators or storytellers with whom one identifies (Mackenzie 1976, Lovell 1998).

The struggle for identities in an unstable world is a result of the breakdown of the relatively unreflexive and totalistic identity schemes in which everyone "knows their place." Increased communication across ever-widening distances and the collapse of custom-based communities disrupt conventional identifications. Yet it is remarkable that even in this world one can say, "Those who share a place share an identity" (Mackenzie 1976, 130). This is so for a number of reasons. First, even as people strive politically to establish identities that are not necessarily place-specific, they do so within a "geographical field" of shared relevance, such as the territory of a state (Calhoun 1994, 25). Second, as they struggle for one identity, people usually share other identities, of which the most important are usually those of the people among whom they live. People have multiple identities and loyalties that derive from the overlapping social worlds in which they live their lives (Calhoun 1994, 26; Sen 2006). Third, communication, social interaction, and reactions to distant events are all filtered through the routines and experiences of everyday life. For most people these are still geographically constrained. Even if not always strictly localized, "shared social spaces" still define the limits for the social appropriateness of given identities (Mackenzie 1976, 131). Finally, "imagined geographies" are important within many identities, such as that of African Americans with its roots in African diaspora, slavery, the southern U.S., and contemporary expression in the "black ghetto," or migrant identities of diasporic groups caught between different social worlds (Mohanty 1991; Davies 1994; Morley and Robins 1995). Places are thus shared, if only in the imagination.

The concept of identity is not without its problems (Detienne 2003). The term itself can seem to imply a solidity and permanence to "identities" that the politics of identities is all about establishing. There would be no point to the politics if the identities were already secure and not subject to denigration by more powerful "others." The danger of "essentializing" identities does not mean that identities are socially constructed "out of thin air" without any meaningful relationship to "natural differences." Given that much racial discrimination is based on reactions to differences such as skin color, for example, it is not surprising that an "African American" identity would involve reference to "blackness." This does not mean that there are the sorts of *real* racial differences of bundled biological and cultural traits that were the stock-in-trade of classical political geography and early twentieth-century "racial science." Genetic research has shown that human DNA is surprisingly similar even among the most genetically "distant" ethnic groups such as the Pygmies of subtropical Africa and the Maoris of New Zealand. So-called racial traits are simply the product of long-term adaptation strategies of the human organisms to different environments and climates. It means only that such differences are *seen* as defining social and cultural differences that feed into identities. In this case the history and expectation of discrimination is more important than anything else. Finally, the relationship between identities and interests is ambiguous. The theoretical problem lies in drawing too neat a line between the two, as if identities and interests must stand in opposition to one another. In fact, political mobilization and action in any particular case cannot usually be reduced to either one or the other.

Even given the generally postmodern tenor of the concept of identity, in particular its constitution out of stories rather than psychological traits or economic interests, the relationship of geography to the politics of identities has been addressed from within each of the three broad streams of political-geographic thinking. Each type of perspective, however, tends to pick up on a specific aspect of the relationship. From a spatial-analytic point of view the focus is on the *boundaries* (social and jurisdictional) that help to define political identities. Political-economic approaches are more interested in the processes of *spatial inclusion and exclusion* that help to create the circumstances in which groups can acquire identities. Postmodern perspectives, broadly construed, privilege the ways in which identities are expressed through attempts at *associating identities with places*.

Alexander B. Murphy (1993) provides an example of the approach that emphasizes the causal connection between boundary delimitation and identity maintenance/formation. Jurisdictional boundaries within states, in this case that between Dutch-speaking (Flemish) northern Belgium and the French-speaking south, are seen as boosting the identities of the groups who already inhabit the different regions. As the pursuit of "ethnic identity," particularly by the historically less powerful Flemish, has come to dominate Belgian politics, one solution has been to devolve government functions to the regional level, thus reinforcing the differences between regions and the identities of the two groups. Regional boundaries within Belgium, therefore, have reinforced if not offered a new foundation for the social identities with which Belgian politics as a whole has become increasingly intertwined. From this perspective, boundaries can be read as signifying the mutual acceptance of different zones of interaction and practice for social groups within one country.

But from a political-economic viewpoint, this approach remains agnostic about whose interests have been best served and what the logic behind the need to define such rigid boundaries has been. Rather than focus on the boundaries themselves, therefore, a more appropriate concern should be with the processes of spatial inclusion and exclusion that boundary delimitation represents. This is the approach taken by Löic J. D. Wacquant (1994) in his account of the contemporary black ghettos of large American cities. Though not politically divided from central cities, these are effectively separate social worlds in which quite different social and economic processes prevail from those in surrounding areas. Thirty years ago black ghettos, such as Harlem in New York, the Southside of Chicago, and South Central in Los Angeles, were also separated. But the new "hyperghettos" have lost the mixture of social classes and have different relationships with the wider society than was formerly the case. True to a political-economic perspective, Wacquant sees pressures from the wider society as crucial in this transformation, with internal changes having only indirect impacts. Using Chicago as his illustrative case, Wacquant shows that the "new" ghetto has two distinctive features: (1) a decaying inner core with satellite working-class and middle-class neighborhoods, and (2) a massive amount of physical decay and social collapse in the ghetto core. The "classic ghetto" gained its communal strength and identity from an organizational infrastructure (churches, lodges, the black press, etc.) that has withered away. Abandoned by government and industry, the hyperghetto is a wild zone beyond ordinary society, lacking in legal job opportunities, good schools, and much hope for the future.

This narrative of a "spoiled identity" suggests the central dilemma for identity politics from a political-economic perspective: that perpetuating an identity, in this case that of the contemporary black ghetto and its people, may mean abandoning the possibility of pursuing constructive interests. These can only be pursued elsewhere; hence the emptying out of the core ghetto by its most upwardly mobile residents. David Harvey (1993, 64) clearly identifies the tension between identity and interests for the political-economic perspective when he writes:

> The identity of the homeless person (or the racially oppressed) is vital to their sense of selfhood. Perpetuation of that sense of self and of identity may depend on perpetuation of the processes which gave rise to it. . . . [T]he mere pursuit of identity politics as an end in itself (rather than as a fundamental struggle to break with an identity which internalizes oppression) may serve to perpetuate rather than challenge the persistence of those processes which gave rise to those identities in the first place.

A seemingly happier coincidence between identity and interests is mapped by Benjamin Forest (1995) for a group of gay men who used the process of creating a new municipality in West Hollywood, California, in 1984 as a strategy for destigmatizing and gaining popular acceptance for their identity. Forest surveys the coverage given by the gay press to the incorporation of a new local government jurisdiction and identifies from their accounts a set of themes that related "gayness" to the place that was being incorporated. As a "gay place," West Hollywood became a concrete anchor for an abstract identity that was otherwise both intangible and threatening to the larger population. A gay identity was thus stabilized and advertised as nonthreatening by associating it with the achievement of political independence for a hitherto unincorporated area of Los Angeles County, wedged between the City of Los Angeles and Beverly Hills.

The place was crucial to this campaign because it allowed a set of usually separate features of gayness to be brought together, both stereotypical (creativity, aesthetic sensitivity, etc.) and counter to stereotypes (maturity, commitment, etc.) A moral narrative was constructed that used the relatively high percentage of gay men in a place as a positive rather than a negative phenomenon. This case is taken as an illustration of the more general claim about how representations of places figure in the making and remaking of identities, a major theme of postmodern and postcolonial thinking in political geography (e.g., Duncan and Ley 1993; Thrift and Pile 1995; Nuttall, Darian-Smith, and Gunnar 1996).

Whether identity politics involves race, ethnicity, or sexual orientation, therefore, there are important links to the places in which the identities in question are defined and pursued. What one makes of identity politics and how one construes the role of geography in how it operates, however, will depend on the theoretical perspective that is adopted. Like the identities themselves, the meanings ascribed to the role of geography in identity politics are highly contested.

Geographies of Nationalism and Ethnic Conflict

The word *nationalism* dates from the late eighteenth century, although there is considerable dispute over whether it is a totally modern phenomenon or has older roots. The multiple forms that nationalism has taken make it hard to offer definitions that are all-encompassing. This has not discouraged attempts at doing so. Two approaches have dominated. The first views nationalism as a political ideology that exalts the "nation" to a central value and in which "national interests" supersede all others (Breuilly 1982; Hobsbawm 1990). The second sees nationalism as an autonomous social force or causal variable in history that, arising first in England or Germany, spread through a dual process of elite imitation of existing "models" and mass disaffection with existing identities first into the rest of Europe and then around the world. The element of attachment to a people or *volk* goes back to the German philosopher Hegel, but the emphasis on elite imitation is a more recent innovation (Greenfeld 1992).

If the problem with the first approach is that it dissolves nationalism into "its" particular manifestations and sees it as derivative of the rise of modernity, the problem with the second is that it reifies the "people" as a primordial entity and regards nationalism as a natural inheritance from the past rather than something that must be constantly recycled and reworked to keep it fresh. Each approach misses one or more key feature of nationalism that the other identifies. On the one hand, nationalism is a type of practical politics mobilizing groups by appealing to common identities and interests. But it is also, on the other hand, a set of ideas about the "nation" as the singular reference for an identity that began as the vesting of sovereignty in the people (as in France and the United States) and then spread under the label of "national self-determination" elsewhere largely on ethnic grounds. In other words, if the politics of nationalism concern the pursuit of presumed shared interests and identity as a population occupying a common territory, the sense of communality that justifies this rests on the appeal to a mythic national past.

From this point of view nationalism is a type of politics that depends on claiming a nonpolitical legitimacy to gain control over a state and pursue other goals, such as expanding the national territory or excluding those who are not thought to share the common past. The relationship to statehood is crucial. Only with the rise of the territorial state does the nation appear as the natural and unitary form to which the state must aspire. The idea of the "nation-state" at the heart of nationalism presupposes the creation of either cultural uniformity (as in German-style ethnic nationalism) or a civic "religion" based on a founding myth and a set of "special" institutions (as in American-style civic nationalism).

In recent years, particularly with the end of the Cold War and the discipline this imposed on both international relations and the possible "break-up" of existing states, nationalism has been strongly associated with the proliferation of ethnic conflicts, even though many of the conflicts to which the qualifier "ethnic" is applied also have other roots (caste, class, and region, among others). Many states are multiethnic and thus subject to this pressure. The bureaucratization and corruption of existing states, regional economic disparities, and the unfreezing of Cold War boundaries have contributed to the upsurge in communal and ethnic conflicts. These are particularly intractable when they involve competing territorial claims, few alternative identities, and interethnic economic competition. Violence is important for validating the seriousness of your claim and encouraging others to accept your interethnic conflict as a zero-sum or winner-take-all game. Without the protection of your own state, there are no guarantees against the violence directed at you by the other groups (Agnew 1989, Kaufmann 1998).

Yet states have increasingly begun to accommodate ethnic and regionalist movements preaching secession or regional autonomy. Often movements will settle for much less than outright separation, satisfied with regional devolution, language rights, and recognition of their "difference." Some scholars have seen the number of ethnic (and related) conflicts as declining (Gurr 2000). Even if some of the 1990s ethnic conflicts may have found resolution, as in Bosnia, Kosovo, Papua New Guinea, or Sri Lanka, a number of high-profile, seemingly intractable conflicts, such as that between Israel and Palestine, or those in the Russian Caucasus, Nigeria, Baluchistan, and Burma, suggest that it is necessary to be careful before jumping to conclusions. As long as there are multinational states and distinctive ethnic identities, the tensions between the two can degenerate whenever one group's identity is viewed as oppressing the others. What is also clear, however, is that many ethnic groups manage to cooperate with one another without large-scale violence (Fearon and Laitin 1996).

The question of how national (and ethnic) groups establish and maintain their territorial boundaries is a contentious one. It is usually addressed indirectly within the confines of the two broad approaches to nationalism outlined previously: instrumentalism and primordialism. Daniele Conversi (1995) provides a spatial-analytic alternative to these nongeographical approaches. He does so by drawing together three theories—the ethno-symbolist, transactionalist, and homeostatic—around the question of boundary definition. He sees nationalism as resting on a process of social categorization in which groups identify themselves in opposition to other groups, with preexisting ethnic markers serving to differentiate them from one another. Internal "ethnic" content and the spatial segmentation produced by territorial boundaries thus interact to provide the basis for a particular nationalism. One may rely more on content, another more on opposition or antagonism across boundaries. In either case, "nationalism is a struggle over the definition of spatial boundaries, that is, over the control of a particular land or soil" (Conversi 1995, 329). Many recent conflicts given the ethnic label seem to fit well into this categorization, and much spatial-analytic research has been devoted to them (e.g., O'Loughlin and Raleigh 2008; O'Loughlin and Witmer 2011).

Conversi is concerned with how boundaries fix nationalism. But he pays no attention to the material conditions under which nationalisms of various types emerge and flourish. Colin H. Williams (1989) uses a version of world-systems theory to frame discussion of the explosion of ethnic separatist movements around the world since the 1970s. He uses case studies of separatist nationalism from Spain, France, and Nigeria to make the claim that the resurgence of nationalism in its separatist guise is the "playing out of minority aspirations unsatisfied during the critical period of state formation" (C. H. Williams 1989, 340). Changes in the political economy of capitalism, particularly the renewed attractiveness of small areal size with globalization, now make it feasible for those groups ill digested by existing states to strike out on their own. A range of local factors also figure in each case—for example, the aftermath of the Civil War of 1936–1940 in Spain for Basque nationalism, or the hypercentralization of the French state for Corsican nationalism—suggesting that the material determinants cannot be solely responsible. Other political-economic perspectives would be likely to give greater emphasis to the degree of forced cultural assimilation, economic exploitation, and military suppression of subsidiary groups.

Many scholars have become disillusioned with the typologies (ethnic versus civic, separatist versus chauvinist, integral [fascist] versus unification, etc.) and lists of conditions that both spatial-analytic and political-economic perspectives tend to come down to. How the national or ethnic past from which the present identity draws is remembered has become increasingly important for these critics. How is the claim to primordiality reworked and reinvented to keep people in the fold? In this postmodern construction, the production and configuration of national images in film, literature, landscapes, and monuments are viewed as the ways in which national identities are forged and reshaped.

The sacrifices of war and the heroism of national political activists are the most frequent subjects of monumental commemoration. Nuala Johnson's (1995) article examines the role of such monuments in Irish national identity. She argues that landscape forms such as statues not only are constant reminders of a collective past but also help to "spatialize" public memory by linking the history of the nation to specific, concrete sites within it. This serves to remind everyone, however remote from the actual battleground of nationhood, of communality with their fellow co-nationals. The nation itself can also be represented in statue form, frequently as a female: a heroic maiden or doting mother signifying the "land" for which so much has been sacrificed. Johnson points out, however, that the meaning of statues is not straightforward, even when evidently celebratory or heavily gendered. They are subject to contending interpretations as to what they say about the past. Hence, there are always possible reinterpretations that can lead to new views of the past. What is clear is that, from this point of view, nationalism is never finally written in stone even if the statues themselves are.

Understandings of nationalism and ethnic conflicts differ considerably across the three sets of perspectives. What is becoming obvious, however, is that rather than being totally antagonistic in their assumptions and approaches, the three offer possible complementarities, given their different orientations and distinctive voices. That they still remain separate perspectives speaks to the continuing impact of affiliations to different traditions of thought and the intellectual tribalism (training, journal policies, habits of mind, reading lists, etc.) that keeps the field often divided more than engaged in common conversation.

CROSSING THEORETICAL DIVIDES

There are signs of theoretical rapprochement. This is particularly the case in some recent research and writing on geopolitics and the rise of deterritorialized forms of power, such as those associated with so-called global cities and world-city networks. For example, Agnew and Corbridge (1995) attempt to bridge the divide between political-economic and postmodern perspectives by showing how "geopolitical order"—based on trends in the practical political economy of world politics—and "geopolitical discourse"—the ways of seeing and thinking about world politics—interrelate to produce the everyday practices of world politics. They identify different historical periods in which order and discourse relate to one another in distinctive ways to resolve the difficulty that arises in ahistorical accounts that give priority to one over the other. Their historical geopolitics represents an attempt at engaging with both political-economic and postmodern perspectives rather than privileging only one of them. Similar, if less formal, attempts at relating political-economic to cultural aspects of geopolitics also animate such feminist writing as that of Cynthia Enloe (1990) and the writing on global cities of Saskia Sassen (1991).

In a different vein, an interesting combination of political-economic and spatial-analytic perspectives is apparent in the Beaverstock, Smith, and Taylor (2000) project on world-city networks. Criticizing the state-oriented understanding of the world that dominates both world politics and the social sciences, they propose instead a focus on the world of flows, linkages, and connections among the world's cities. They bring together the explosive growth of service industries, the increased importance of information technology, and the tremendous development of worldwide direct investment to propose a theoretical framework for mapping what they call "the intercity global network." Taking a roster of fifty-five world cities, they show how firms from some set up shop in the others (figure 4.7). Empirical analysis of these linkages shows that there is a definite hierarchy to them, with some cities, such as London and New York, currently sitting on top. This pattern suggests that "networks" are beginning to challenge "territories," if not for

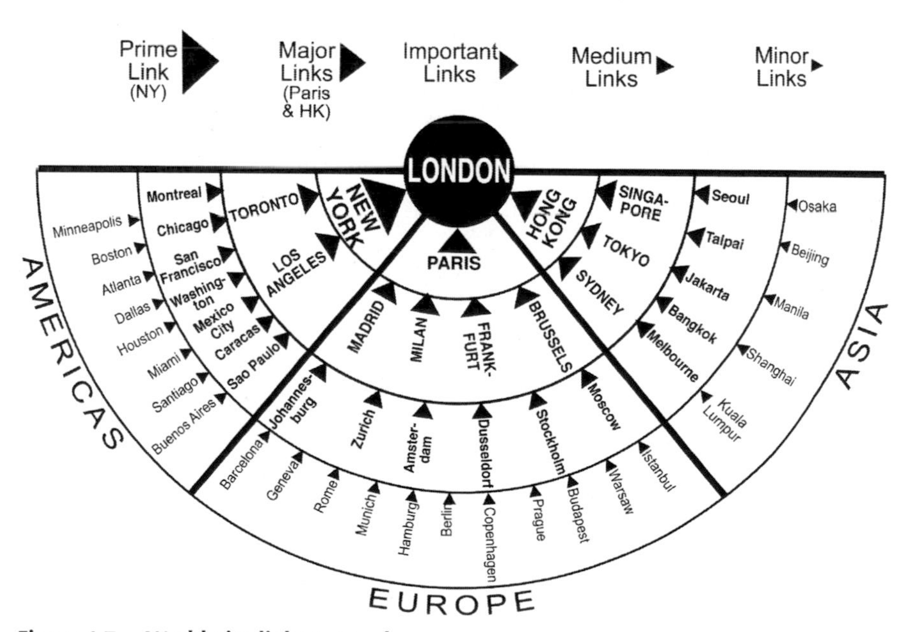

Figure 4.7. World-city links to London.
Source: Beaverstock, Smith, and Taylor (2000)

the first time (see, e.g., Rokkan 1980 on the Central European city belt), as an organizing principle of global geopolitics. Importantly, however, the authors note that "world cities are not eliminating the power of states; they are a part of a global restructuring which is 'rescaling' power relations, in which states will change and adapt as they have done many times in previous restructurings" (Beaverstock, Smith, and Taylor 2000, 132).

In the study of nationalism and ethnic conflict, there is a different rapprochement between spatial analysis on the one hand, emphasizing the spatial patterning of conflicts, and on the other hand postmodern approaches to interpreting the narratives or stories told by the combatants. Good examples come from the joint research of John O'Loughlin and Gearoid Ó Tuathail on ethnic conflicts in Bosnia and the Caucasus (e.g., Ó Tuathail and O'Loughlin 2009; O'Loughlin and Ó Tuathail 2009) and Gerard Toal and Carl Dahlman (2011) on the return of displaced people to Bosnia in the aftermath of conflict. The latter study is an exemplary attempt at weaving together what can be termed "facts on the ground" about the geography of ethnic "cleansing" and the political debates and different stories of various parties to the resettlement of displaced people, including foreign participants. Such research is not classifiable in terms of a single "approach" and is suggestive of future possibilities beyond the threefold categorization that tended to characterize political geography at least into the 1990s.

In the future we can expect to see more "violation" of the theoretical boundaries that have subdivided political geography in the recent past, if only because the world to which political geography is applied is changing rapidly and in ways that make the previous "intellectual division of labor" increasingly irrelevant. In particular, the emerging "geopolitical order" is one that challenges the fixed spatial claims of spatial-analytic perspectives, the unchanging political-economic imperatives of many political-economic perspectives, and the focus on dominant discursive representations of postmodern approaches. Chapter 5 addresses both the emerging post–Cold War geopolitical order and the new themes (and associated theoretical perspectives) of the political geography that is coming about alongside it.

5

The Horizon

Considering the increased diversity and pace of relevant events shaping world politics since 1990, the very fact that the post–Cold War period hasn't yet found its distinctive name, and continues to be defined as after or beyond the previous period, suggests not only that this is a time of transition from the relatively fixed geopolitics of the bipolar Cold War era to an as-yet-unpredictable one, but also that this period remains at least in part conditioned by continuities that stem directly from the Cold War. As state-centered views of geopolitics have been challenged both by new transnational networks and place-making activities, a number of identities previously trapped in the Cold War bipolar order have been in search of new territorial expression, if often at different geographical scales from those of the Cold War past. Contemporary political geography, therefore, while still tied to past approaches and debates, is being made increasingly in relation to new geopolitical events and trends that are not defined by a single overarching story such as that of either the Cold War or the interimperial rivalry of the early twentieth century, despite attempts at imposing a single narrative such as the "War on Terror" or the "Rise of China."

The growing evidence of the geopolitical relevance of issues relating to globalization, on the one hand, and new place-based and territorial identities, on the other, has generally been recognized in political geography since the end of the Cold War in the early 1990s. Still, one major difference with even the recent past is the complexity of the geopolitical context in which political geography is today being remade as opposed to the relative simplicity of previous periods.

Notwithstanding the evident difficulties in interpreting a period without benefit of hindsight, and with the additional complexity due to the multiplication of relevant actors, media hype, disinformation, and lack of official documents that are likely to remain classified for many more decades, in the following section dedicated to the post–Cold War geopolitical context, we draw attention to what seem to be some of the dominant features of global geopolitics from 1990 to the present, both in the entire period and before and since 2000. We identify these as an emerging interdependent world, identities in search of territories, the War on Terror, increased demands for democracy and self-government, and the crisis of the territorial model of political organization.

THE POST–COLD WAR GEOPOLITICAL CONTEXT

The final collapse of the Cold War system first mapped out at international conferences at Bretton Woods and Yalta in 1944–1945 had actually begun two decades earlier than 1990. These two agreements imposed geopolitical order, respectively, on world monetary and territorial affairs. Bretton Woods created a fairly stable system of fixed currency exchange rates, and Yalta froze the boundaries of postwar Europe and established the spheres of influence of the Cold War. The Bretton Woods Agreement ended in 1971 with the U.S. abrogation of the use of the dollar as the world's reserve currency and of the U.S. as the lender of last resort. In 1989 Yalta came to a close. The end of the bipolar order started at the urban scale with the downing of the Berlin Wall (1989). It soon extended to the European scale with the reunification of Germany (1990), the collapse of the Iron Curtain, and the disintegration of Yugoslavia, and reached its peak at the global scale with the implosion of the Soviet Union (1991), officially ending the Cold War in 1992.

Compared to previous world epochs, the Cold War era had its own distinctive character. Despite the indirect military confrontation between the U.S. and the Soviet Union in a number of "low intensity" conflicts that took place in the "Third World" (most notably in Vietnam and Central America), the advent of nuclear weapons meant territorial security derived from the paralysis of the most powerful states rather than from the open exercise of their power, and from the acceptance of the impossibility of global monopolization of territorial violence rather than from its pursuit (Deudney 1995). Nonetheless, the Cold War was deeply territorialized. Both superpowers harnessed science and technology to serve a military competition for territorial control that extended to the oceans, the atmosphere, and outer space. This expanded territorial projection—characterized by a bipolar division of the world that appeared at different scales (from global spheres of influence to the division of the city of Berlin) and was reproduced within different world regions and in the new arenas made accessible by technology (such as outer space)—was the product of the deep ideological opposition permeating both sides. This was so powerful that it resembled more a war between "religions," based on such oppositions as West and East, Good and Evil, Capitalism and Communism, rather than a simple old-fashioned clash of empires. A case could perhaps be made that the prevalence of such oppositional logic projected to a degree worldwide was the outcome of a shared psychological inability to cope with the increased complexity of the world, especially after World War II and with the invention of nuclear weapons. The demographic boom that had accompanied the spread of the Industrial Revolution and the reduced possibilities of isolation in view of technological developments and increased world trade and investment only added to the greater complexity. What has replaced the binary logic of the Cold War order? Some of its main features are outlined before turning to the "new themes" in political geography that in many ways reflect its course.

One World Emerging

Still, considering the long psychological cross-pressure involved in the Cold War, the conspicuous financial resources poured into it, and the degree to which its discourse shaped and mobilized the respective societies (including the Western antiwar protests of the 1960s and 1970s), the implosion of the Soviet Union was immediately greeted by the U.S. and its allies as a victory for American ideology, implying the superiority of its values and political and economic institutions (e.g., Dalby 2010). It mattered little that the Soviet "defeat" had been largely self-inflicted rather than a direct outcome of American military action. As the former enemy had to "convert" to the Western political and economic system, the very destiny of the world seemed inevitably inscribed in an extension of the virtues of American democracy and liberal capitalism to the whole globe.

"Globalization" arguably began five centuries earlier, as a western practice of Anglo-Dutch origin that emerged from the maritime trading world (e.g., Blouet 2001), but the state-based territorialized vision of world politics typical of the 1875–1945 interimperial rivalry hampered its development (Agnew 2009). In the context of the Cold War, the U.S. had resorted to globalization in order to ideologically promote a "free world economy" within its sphere of influence. In its contest with the Soviet Union, from the late 1940s on, America presented itself to the world as an agent of economic rationality based in private ownership and the spread of private consumption that contrasted with the collective conformity and limited personal freedom of Soviet life (Agnew 2009, 17). Indeed, this was coherent with the project of taking advantage of foreign opportunities for its businesses, an idea that goes back to the failed post–World War I plan for U.S. economic expansion beyond the colonial purview of the European empires (e.g., N. Smith 2003). In the ideological context of the Cold War's propaganda, its operation became possible thanks to the diffusion of time-space compression technologies, many resulting from military research. Therefore, even before the 1980s spread of a neoliberal ideology that explicitly promoted economic globalization, "markets" had already begun to acquire powers heretofore vested in leading states. As the process intensified and expanded, some localities and regions within states were privileged within global networks of finance, manufacturing, and cultural production; New York, London, and Tokyo rose to the status of global financial centers.

Differently from the aftermath of the previous world war, this time the "victors" had no ready plan for a new world order, no blueprint for peace, no plan for the economic reconstruction of the vanquished. In fact, it even seemed that there was no need for one. As the U.S. journalist Charles Krauthammer (1990) was announcing America's "unipolar moment," the unexpected disappearance of the enemy had suddenly left the U.S. as the sole superpower. While Luttwak (1990) theorized the emergence of a new world of geoeconomics, where the "grammar of commerce" would replace military methods as the primary "tool of statecraft" and become the main instrument of a new foreign policy and international relations "without regards to frontiers," the end of the Cold War was greeted by a blooming of "endist" theories, declaring the end of history (Fukuyama 1992), the end of geography (O'Brien 1992), the end of the nation-state (Ohmae 1995), and the end of territories (Badie 1995), as if a U.S.-centered globalization was bringing about the total eclipse of geopolitics. And then the first President Bush announced his "new world order," whereby the global role of guaranteeing peace, security, and freedom was to be undertaken by the U.S.-UN couplet, soon to be seen in Kuwait, Bosnia, and Somalia.

In the euphoria of victory, an ideological equation was therefore established between the United States and globalization, seen as the inevitable Americanization of the world. American identity could now be projected onto the whole earth, while a world of opportunities had finally opened up even beyond the former Iron Curtain. Increasingly seen as a substitute for politics, the "market-access" regime of globalization would tie local areas directly into global markets. Successful ones could enhance their position by increasing their attractiveness to multinational and global firms. Unsuccessful ones would be isolated from the process. But that mattered little to the "victors" of the Cold War: the ideological equation went even further to the point of naturalizing the whole process, as if American globalization represented the singular answer to the question of the very destiny of the human species (e.g., Caracciolo 2011). The emerging world that "knew no borders" would be an American one. That was the final outcome of the Cold War. But, as we shall see, it did not stay that way for long.

Identities in Search of Territories

Divided into two spheres of influence, Europe had been the Cold War's central front. To the East, the geography of territories frozen by the Cold War had already started to melt under the pressure of social and political movements. The opposition between communism and capitalism had erased or obscured every other preexisting rivalry or conflict (Hobsbawm 1994), and when centralized Soviet authority collapsed, ethnic tensions exploded. Impoverished by the collapse of the Soviet economic sphere and long oppressed by Moscow's totalitarian state, ethnic identities, based on the national self-determination principle, began to reverse not only the Cold War's territorial order but in some cases even what remained of the one agreed at the Paris Peace Conference of 1919. Seven states were to be born out of Yugoslavia; the Czechs consensually divorced from the Slovaks, and out of the Soviet Union fifteen states became independent. The tidal wave of new states followed the cultural fault lines across the Eurasian heartland but left many groups dissatisfied. In the Caucasus the problem of Chechen and South Ossetian identities was left unresolved. The coherence of Russian identity was questioned within its borders, as shown by demographic indicators suggesting a lack of faith in the future (declining birth rates, increased death rates, and declining life expectancy). By the mid-1990s birth rates had dropped sharply and male mortality peaked largely as a consequence of alcoholism, to the point of creating a massive gender imbalance.

The ethnically mixed Balkans saw a number of identity-based conflicts in the early 1990s (Thual 1995; Dijkink 1996). First Croats and Serbs clashed, then war exploded in Bosnia, where ethnic and religious identities were especially intermingled. Even in the postwar period a struggle between the national separatists and the international community continued where the Dayton Peace Accord had left unresolved territorial issues (Dahlman and Ó Tuathail 2006, Toal and Dahlman 2011). In Kosovo, the roots of the conflict triggered by ethnic Albanians' struggle for independence were identity-based on both sides, as the Serbs saw the land as the historical "cradle" of their national community even though they were now a tiny minority of its population (Klemencic 1999).

The governments of Western Europe were unprepared to respond. Unable to stop Serb-dominated Yugoslavia's ethnic cleansing in Bosnia or to agree to a shared political line among member countries, Europe had to ask for a UN-U.S. peace force to "mediate" in Sarajevo. Beyond elite discussion, there was effectively an ambiguity in European identity: were the Balkans inside or outside of Europe (Balibar 2002)?

To the U.S. government, the collapse of the Soviet Union meant that the European "problem" had finally been fixed: Western Europe was no longer a strategic priority. In the shared ideological euphoria of the "victory," however, it could not immediately effect such change in practice and kept thinking of itself as if the Cold War order was still in place, at least from the point of view of continuing American military protection abroad. Fifty years after its birth, NATO had its baptism of fire as an alliance with the "humanitarian war" waged against Milosevic's Yugoslavia to protect Kosovo Albanians.

The memory of two world wars and related totalitarian regimes, the Iron Curtain, and forty years of "Americanization" had not allowed for the development of a unified Western European political identity, while European linguistic policies had actually reinforced the different national identities. If NATO was the military pillar of U.S. Cold War strategy in Europe, the political-economic pillar had been the American-supported European integration process, embodied since 1957 by the European Economic Community (EEC). The furthering of the already planned economic integration process was seen as the best path toward the goal of a unified European political identity, especially after the reunification of Germany. The EEC morphed into the European Union in 1993, internal borders were abolished (1995), and most national currencies switched to the euro (2002), while the enlargement of its membership went from twelve states in 1991 to twenty-seven in 2007, mostly through the addition of Eastern European countries. But a shared political identity remained absent.

In Africa, the end of the bipolar order added more tragedies to those already produced by European colonization and by a decolonization process that too often was "a gigantic footnote in the Cold War" (Gallagher 1982). Its impact was direct in Somalia, Ethiopia, and Zaire/Congo (Kinshasa), which at different times had enjoyed Soviet support. A less direct effect was that related to the loss of Africa's strategic significance in the "new world order": civil wars erupted in Rwanda (with the tragic genocide of eight hundred thousand), Angola, Liberia, Sierra Leone, Ethiopia-Eritrea, and Sudan, among other conflicts. It is actually hard to find an African country that was not affected by the Cold War and therefore by its ending. Even when control of strategic resources was at stake (as with oil in Nigeria), a major role at the national and subnational scales was played by ethnic and, sometimes, religious groups clashing over whose identities would prevail in a given territory. In this historical context, perhaps one of the few bits of good news was South Africa's abolition of its apartheid or racial separation regime in 1994.

In Asia, the end of the bipolar order could be seen in British Hong Kong's reversal in 1997 to Beijing's control and the 1951 border between Hong Kong and China being changed into a Tijuana-style (U.S.-Mexico) immigration barrier. Also, the Cold War border between the two Koreas came under increasing pressure in the 2000s, as North Korea started to develop nuclear weapons. Though it is actually more an outcome of a long-delayed decolonization process, in the context of the U.S.-UN "new world order" the United Nations played a major role in the independence of East Timor (2002) from Indonesia.

The Middle East remained a U.S. strategic priority due to its longstanding support for Israel and the supposedly increased need to maintain a stable oil supply after the loss of Iran as an ally. Saddam Hussein's invasion of Kuwait (1990) provoked the First Gulf War against Iraq (1991), accompanied by the militarization of the Persian Gulf. In Saudi Arabia, as the home of Islam's most sacred sites, the U.S. presence was strongly resented, fueling anti-American radicalism around the Muslim world. In reaction to the sense of a traditional identity threatened by American modernization, Iran had already erupted in revolution (1979), where the Muslim Shiite priesthood established a theocracy under Islamic law. In the following decade, as Islamic radicalism continued to spread across the region, it was increasingly exploited by the Reagan administration in the Cold War. The U.S. CIA trained and armed the so-called "Arab-Afghani" guerrillas, gathered from a variety of Muslim countries, to fight against the Soviet occupation of Afghanistan in 1979. After the withdrawal of Soviet troops (1988), the civil war continued for the following decade, as the Northern Alliance (largely ethnically Tajik) fought the Pakistani-sponsored Taliban (largely ethnically Pashtun). With the Soviet defeat, Afghanistan lost its strategic interest to America, leading the CIA to drop its support for the Arab-Afghanis. As the Saudi monarchy had allowed the U.S. to militarize the Persian Gulf, the anti-American resentment of the Islamist warriors, already upset at U.S. support of Israel, quickly turned to rage (Engel 2001). America's change in strategic interest had also affected neighboring Pakistan, which gained its attention only when its confrontation with India—a conflict that went back to the partition of British India in 1947 and had already caused two wars—peaked as both countries tested their nuclear weapons (1998).

Among the signs announcing that the U.S. had become the target of Muslim jihadists or holy warriors would be the failed attack on the World Trade Center (1993); the bombings of U.S. embassies in Nairobi, Kenya, and Dar es Salaam, Tanzania (1998); and the suicide attack on the USS *Cole* in the Yemeni port of Aden (2000). As the 9/11 Commission (National Commission 2004, 259) has reconstructed the run-up to the later terrorist attacks in New York City and Washington, "the system was blinking red." Still, eager to relax after the long Cold War, reassured by the unchallenged superiority of the U.S. military machine, distracted by the rhetoric of American globalization, the U.S. administration failed to recognize the intelligence warning as a priority. In its first eight months, the administration of the second President Bush had already outlined a foreign policy echoing isolationist tendencies, visible when the Congress voted for the withdrawal of U.S. troops from Somalia in 1995, after the "Black Hawk Down" incident. America was then expressing its own aporia: the paradox of a country caught between the necessities of an unstoppable globalization requiring global military responsibilities and a "private Idaho" desire for the return to an impossible isolationism, echoing a nostalgia for a nineteenth-century world politics based on entirely separable territorial states. This aporia resulted in a tragic overlooking of those identity conflicts about Islam and Americanization that had long been burning within the Islamic world.

The War on Terror: America Fighting Itself?

The selection of the World Trade Center in New York City and the Pentagon outside of Washington, DC, as terrorist targets located in global economic and political centers had a deep symbolic significance. The twin towers were a global symbol of New York City and the world economy, especially as a seat of global finance. As the seat of the U.S. defense department, and in the context of its unrivaled global military superiority, the symbolism of the attack on the Pentagon was understood by American political leaders as perhaps even more disquieting. Inspired by a hatred of "America" as a geopolitical abstraction in a fanatic religious cosmology associating the United States with the personage of Satan, the terrorists understood both the symbolic significance of where and how they struck and the best geographical strategy to achieve their goal. In doing so they challenged the established vision of world politics as entirely international, that is, that conflict is always between identifiable territorial states. As it was soon learned, the terrorists operated as a shadowy network of like-minded individuals with territorial bases in Afghanistan or Sudan and so-called sleeper cells in American and European cities. Their financing was channeled through supporters located in a number of states, including Saudi Arabia and other "conservative" Gulf states.

On the morning of September 11, 2001, America awoke to a nightmare that surpassed any disaster movie's possible imagination. The news media—already in place to cover the first strike—broadcast live to millions the second airplane crash into the second tower. As both towers collapsed in front of the eyes of the world and images of the tragedy endlessly looped on the screens, the shock of the attacks went global. In America, it felt like a veil had been broken. The country's identity had been questioned in depth. The reaction that followed can be understood using Gottmann's approach contrasting opportunity with security (see chapter 4). Right after 9/11 the demand for security immediately affected long-assumed ready accessibility to all places at all times. All roads and bridges into Manhattan were blocked. The U.S.-Mexican border was placed on highest alert. U.S. airspace was shut down, and all civilian aviation grounded for three days. As requests for information paralyzed news corporations' web servers and phone networks, the NASDAQ, New York, and American stock exchanges closed for a week. Globalization seemed to have been stopped. Moreover, America reacted to the identity crisis by reinforcing its national iconography, the most evident signs of which were American flags flying all over the country. Episodes of intolerance targeting anyone with a Middle Eastern look were reported. Afraid of being mistaken for Arabs, Jewish people in New York City feared leaving their homes. An Indian American was shot to death in Arizona after having been mistaken for a Muslim because of his Sikh turban.

The Manichaean cosmogony of bin Laden and his terrorist network, associating America with evil, produced a return to the Cold War's bipolar thinking, as actually many in the second Bush administration had already been "cold war warriors" accustomed to thinking in oppositional terms by decades of such experience (the "need" for an enemy eventually was to bring some of these to later replace Islamist terrorism with China as the primary threat). The government framed the terrorist attacks as an act of war that required a military response. This was to be articulated in terms of three pillars (Caracciolo 2011): First, in order to project America's fear beyond U.S. borders, terrorism's global network was to be fought anywhere, for an indefinite time, and with all possible means (Gregory 2011). As an asymmetric threat, it also required electronic surveillance, monitoring of financial transactions, ethnic profiling, biometric border controls, and, as was to be seen, "extraordinary renditions" and "enhanced interrogation techniques" (torture). From the evening of the same tragic day, the enemy had the face of bin Laden and his al-Qaeda organization (named after an Arab-Afghani training camp). Second, to reestablish the international prestige of the U.S. and reassure domestic public opinion, a "global war on terror" was launched. The display of hard power came soon with the invasions of Afghanistan and Iraq, marking a passage from the 1990s' unipolar moment to the 2000s' unilateralism (NATO's offer for assistance was declined), in a regressive solipsism that announced a return to old-style territorial geopolitics. The justification was that of fighting "rogue states," presumed to harbor terrorists and, for Iraq, of having weapons of mass destruction. The representation of these countries as belonging to an "axis of evil" (that also included Iran and North Korea) echoed the language previously used to define the Soviet enemy. The third pillar was an obsession with domestic security and the reconstruction of the United States as an unassailable "fortress," to be addressed through the creation of the Department of Homeland Security and the Patriot Act. In the name of security, fundamental erosion of American civil liberties, international law, and universal human rights was to be justified (R. Jones 2011).

The U.S. War on Terror is a good example of how casting complex problems in terms of national security tends to give priority to the territorial conception of sovereignty over other possible spatial modalities. The war in Iraq was arguably the direct result of confusing a terrorist enemy organized loosely over space (al-Qaeda) with a territorial state (Iraq) and also an outcome of forgetting that the attacks of September 11, 2001, were as much an attack on the current world economy and all its nonterritorial actors (banks, stock exchanges, investors, etc.) as they were an attack on the sovereign territory of the U.S. (Agnew 2009, 107–8). As a result, the first three years after 9/11 resembled a worldwide political nightmare as the U.S. administration's military response was actually met by a dramatic increase in terrorist acts in the "old world." In the name of "national security," the free press, essential to any healthy democracy, had to undergo a number of limitations, from "spin-doctored" information to "embedded" war correspondents. The media often became a device to distort reality, manipulate emotions, and channel public affect (Connolly 2002; Ó Tuathail 2003). And even if the notion of a "conspiratorial state control of the media" has to be rejected, the media role in "promulgating the myth of antagonistic collective identities" is at least responsible for complicity (Jones and Clarke 2006) as the war on terror was turned into a spectacle.

After military "victory" was declared in Iraq (2003), the fact that the U.S. had not prepared to govern a conquered country and was now faced with the difficulty of doing so undermined domestic consensus and led to a questioning of the entire strategy upon which the invasion of Iraq had been based. As the weapons of mass destruction upon which the invasion was premised had not been found, the justification became that of building "democracy." But military conquest alone could not create political stability, let alone a democracy. From 2004 until today terrorism and civil war have turned Iraq into a living nightmare. Moreover, as the Iraq invasion happened in violation of international law, Bush's doctrine of a "preemptive strike" triggered a wave of anti-Americanism even in Europe. This suggested the necessity in some quarters of building a more multilateral foreign policy rather than the "cowboy" style of Bush. At the end of 2006 the secretary of defense, Donald Rumsfeld, was fired, and a new approach to "pacifying" Iraq was introduced with General Petraeus's counterinsurgency strategy, based on a troop "surge" and on a detailed "knowledge of the social terrain of conflict" (Gregory 2008; Packer 2006). Military thinkers now thought they needed to "disaggregate" (Kilcullen 2009), downsize, and differentiate (Ó Tuathail 2010) insurgencies, while specific media campaigned to familiarize the public with the notion that even if the peace could be distant, there was a slow progress toward that end.

The confidence in U.S. military and technological superiority behind the invasions of Afghanistan and Iraq highlighted the extreme political immaturity of its leaders' vision: the unrealistic idea that democracy could be exported with arms was worsened by the essentialization and simplification of the Muslim world as a singular hostile Other (Burke 2011). Existing institutions of global governance were weakened: the already problematic state-based ontology of the UN was delegitimized further by the Iraq invasion. The Iraq war's duration and consequences had been largely underestimated, not to mention the globally damaged reputation of the U.S. as a defender of civic liberties and human rights, in view of the hundreds of thousands detained at one time or another as suspected militants, and the numerous cases of torture and mistreatment (Watson Institute 2011). From one viewpoint, one hundred million Arabs/Muslims were taken hostage for ten years due to the acts of a tiny minority (Al Qassemi 2011).

By not looking at the causes of terrorism, the official U.S. response actually reinforced it and damaged the globalization process in which the U.S. had hitherto invested so much energy. Even if the increased surveillance so far has prevented new attacks on U.S. territory by Islamic jihadis, the number of terror acts has increased exponentially in the rest of the world (see table 5.1, pp. 205–6). A conservative estimate of the human toll of the conflicts in

Iraq, Pakistan, and Afghanistan has calculated a total of 225,000 deaths and 365,000 wounded between 2001 and mid-2011 (including almost 10,000 American military and security contractors). Meanwhile, the terrorist mode of operation has been reproduced all over Europe, Asia, and Africa as suicide attacks multiplied from Madrid to London, from Bali to Moscow. The tragic siege of an elementary school in Beslan, North Ossetia, in 2004 by a group of Chechen separatists fueled by Islamic radicalism, kidnapping over 1,200 and murdering many, provoked the indiscriminate response of Russia's central government, killing almost four hundred hostages in an abortive rescue attempt.

If in the 1990s globalization could be seen as a process largely driven by America, the 9/11 terrorists may well have reached their ideological objective: that of turning America (and the West) against itself. As if the Cold War bipolar logic had been interiorized, the United States became the enemy of itself. According to Caracciolo (2011), the real success of terrorism may have actually been the War on Terror. By overestimating the terrorist threat, instead of addressing other global priorities, the U.S. has spectacularly overlooked the economic rise of China, India, Brazil, and Russia, while the obsession with security restricted the movement of people and trade and capital transactions upon which the globalization process is based. In this way, America may have produced exactly what the terrorists wanted in order to destabilize its hegemony: its global financial and moral weakening. This is a strategic blunder that has both affected the reputation of the U.S. in the world and reduced civil liberties at home. As the national bureaucratic machine was reoriented toward security, a securitization that may remain in place for a long time, little has been done to improve responses to disasters such as Hurricane Katrina or the Deepwater Horizon oil spill in the Gulf of Mexico.

Moreover, the prolonged war effort has been financed almost entirely by borrowing abroad as much as at home, burdening the U.S. public debt. Already at the time of the Iraq invasion, Agnew (2003a) had cautioned about the economic sustainability of the "empire" strategy as a way of preserving U.S. hegemony. A recent study has calculated the costs of the U.S. wars in Iraq, Afghanistan, and Pakistan since 2001 at up to $3.2 to $4 trillion (plus another trillion in additional interest payments to 2020). War costs have not been covered by increased revenues; indeed, the Bush administration lowered marginal tax rates for the wealthy at the same time the wars were under way. Together with deviations within the financial system, already evident with the 2001 Enron corporate fraud and corruption and more recently in the subprime mortgage crisis of 2007–2008, this has contributed to shaking the very foundations of the global financial system.

The mortgage crisis has been explained by Joseph Stiglitz (2008, 2010), the Nobel Prize–winning economist, as the result of "a pattern of dishonesty on the part of the financial institutions and incompetence on the part of policy makers," where the investment community became a victim of the complex financial instruments it had created. The ensuing crisis was fought by the U.S. government with the public bailout of those private banks and companies that had heavy responsibilities in bringing about the crisis itself. A major contribution from China and other countries with healthier economies temporarily prevented a further globalization of the crisis. But U.S. national debt problems, combined with a general sense of uncertainty in the financial markets, renewed the crisis in the summer of 2011, in some cases threatening the very sovereignty of some nation-states as they became subject to policies mandated by international organizations such as the IMF and the European Central Bank.

Changing Strategic Priorities and Demands for Political Change

Meanwhile, the combined effects of the financial crisis, a lack of freedom and democracy worsened by the War on Terror, and the coming of age of a young generation usually representing more than half of the population by the winter of 2010–2011 brought about an unprecedented series of protests in North Africa. Using the full power of digital media (cell phones with cameras, satellite TV, the Internet), hundreds of thousands filled the main cities' squares, asking for freedom, democracy, and a chance at a better economic future and aimed at a radical change away from the authoritarian leaders that had ruled since the Cold War without sharing any economic benefits with the bulk of their people. In a rapid turn of events, the Tunisian and Egyptian governments were overthrown, while the protests, multiplied by digital media, spread throughout the entire Middle East and beyond.

Despite involving mostly Arab people, the so-called Awakening was not a "pan-Arab" movement, since each protest had a distinct national inflection (Filiu 2011). This came largely unexpected in the West, especially because, as some political geographers have pointed out (Falah 2011; Mamadouh 2011), such protests gave a different perception of the Arab people, opposite to the Orientalist one that portrayed a world stuck in a religious past, resistant to democracy and change (e.g., Lewis 1995; Huntington 1993), whose youth could only express itself through Islamic fundamentalism and jihadism. Suddenly, it became evident how ignorant and stereotyped the Western understanding of the Muslim and Arab worlds had been, as the U.S. War on Terror promoted a general Islamophobia that then fueled Islamist radicalism (Reifer 2006). This was exactly the "management of savagery" strategy theorized by al-Qaeda ideologist Naji (2006). Fortunately, despite the fact that the U.S. government fell into the trap, it seems to have failed so far. Instead, here was a new generation asking for freedom and democracy, those same values upon which the West was putatively built, and for a chance at a better future, using social-network media to communicate and organize the protests. And al-Qaeda had nothing to do with it (see figure 5.1).

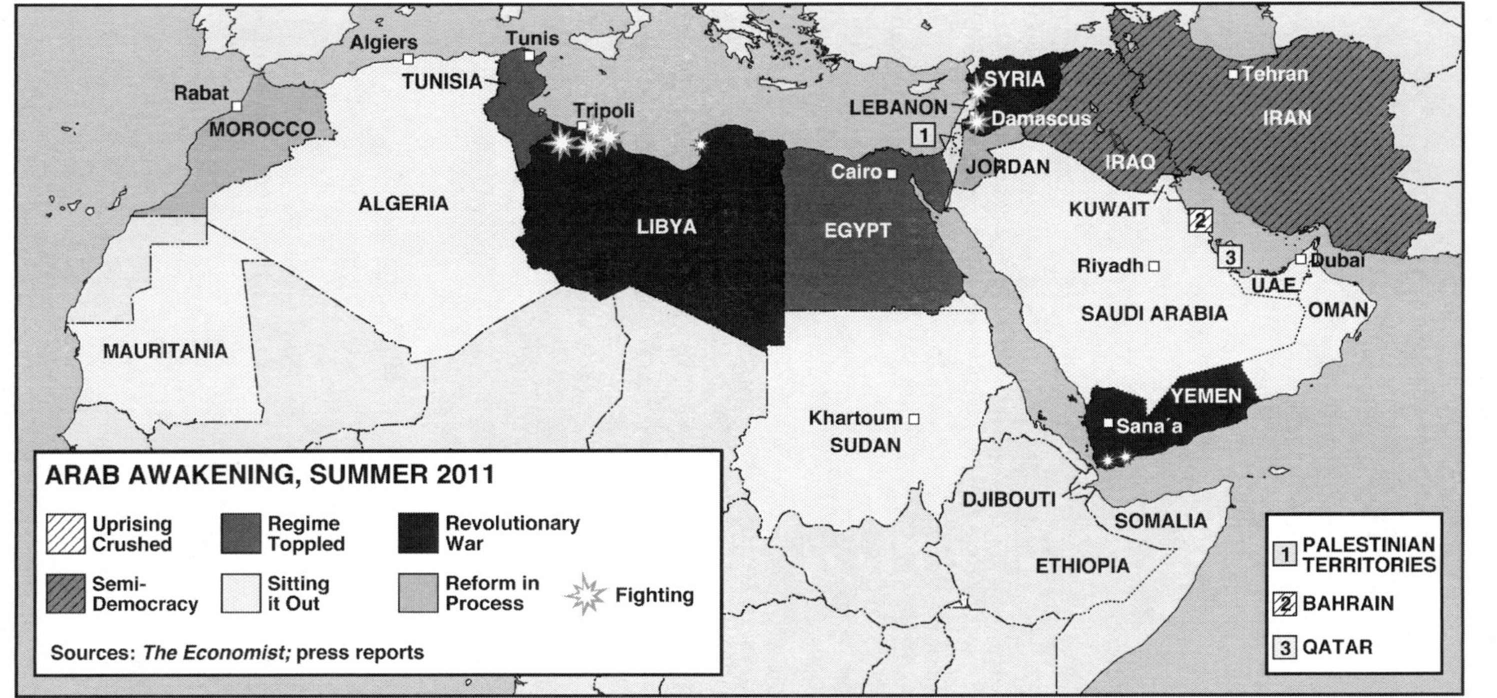

Figure 5.1. The "Arab Awakening" map showing the nature and course of protests in North Africa and the Middle East during winter 2010 and early spring 2011.

Source: Redrawn from a number of sources.

The popular protests spread to Jordan, Algeria, Mauritania, Sudan, Morocco, Yemen, Bahrain, Oman, Iran, Libya, Lebanon, Iraq, Kuwait, and Saudi Arabia, with a potential to destabilize the status quo in the whole of North Africa, the Middle East, and beyond. To be sure, the Saudi king dispensed money and benefits to his population, provided military support to quell protests in Bahrain (rumored to be helped by Saudi Arabia's archenemy, Iran), while sheltering the former rulers of Tunisia and Yemen. In Syria, an ally of Iran, it is open to question if the massive popular protests (the last ones to start) and the bloody repression that ensued signal either a continuation of the trend or that a counterrevolution had begun in the Middle East.

Throughout all this, the United States government remained surprisingly silent until mid-May 2011. As the leadership in Tunisia and Egypt was crumbling, the Obama administration unexpectedly did not move in their support (Egypt being a key ally to the U.S. in relations with Israel). Egyptians were left to proceed on their own path to regime change. Another quite different case is that of oil-rich Libya. When the UN Security Council voted Resolution 1973 in support of rebels against the Gaddafi regime and the war expanded, the U.S. quickly withdrew its bombers and missiles and transferred control to NATO, leaving France (with the UK) to take the lead in what has been called by some (not least the Gaddafi regime) a neocolonial attempt to reestablish Franco-British power in the Mediterranean. Beyond the European confusion this provoked (Germany opposing the war, Russia abstaining, and Turkey trying unsuccessfully to counter it), the U.S. behavior in regard to Egypt and Libya has been questioned, beyond the preelectoral explanations of eschewing military overstretch (Kennedy 1987) and avoiding adding to an already troubled federal budget. If Libya signals an apparent U.S. military disengagement from the Mediterranean, is it a significant turning point in U.S. foreign policy? In choosing U.S. strategic priorities more carefully (e.g., the Pacific and Indian Oceans over the Atlantic/Mediterranean), has President Obama realized that the U.S. cannot afford the role of continuous policing of the globe, or is he simply trying to avoid overextension of U.S. military power in view of the 2012 U.S. presidential election?

Under the doctrine of "smart power" (a mix of soft-power international idealism and hard-power old realism), it may be counterproductive to take the burden of guaranteeing global stability if a number of rising powers end up taking advantage of it. If China, Russia, India, Brazil, and others are challenging U.S. hegemony, as some would see it, the best strategy may be to pull out of those settings that are not defined as really essential, leaving partners and rivals to take the direct burden, in a sort of geopolitical-geostrategic deregulation that announces a return to a balance-of-power framework for world politics. In realizing that the U.S. may rather keep an influence, but not aim at full control over the rest of the world (as if it ever had that), some realist scholars have suggested that Obama may have pragmatically noticed Friedman's (2009) and Luttwak's (2009) advice: to allow a certain degree of anarchy in the international system to keep the challengers busy, to U.S. advantage (Dottori 2011).

Still, the continued unquestioning U.S. support for Israel, irrespective of how the Israeli government behaves, together with the absence of any explicit reference to Saudi Arabia in Obama's May 2011 speech about political change in the Middle East, are enough to suggest little substantive change in the U.S. Middle East policy, even as the costs of twenty years of U.S. militarization of the Persian Gulf are now being questioned (Stern 2010). It remains to be seen if, after the killing of Osama bin Laden, developments in Afghanistan will confirm the hypothesis of a paradigm shift in U.S. foreign policy that would end America's "unipolar moment" both in its globalizing variant of the 1990s and in its War on Terror / Age of Permanent Warfare variant of the 2000s. If the specter of hegemonic decline hovers over the United States' role in the world (Zakaria 2008, 2011), the strategy of "leading from behind" (as Obama's Libya policy has been labeled) may buy time while slowing down the rise of new challengers.

Singularly Territorial Organization of the World in Question

The end of the Cold War brought a revival of general interest in geopolitics, to the point that it has become useful to distinguish three branches: a "think tank" geopolitics (offering advice to policy makers), a popular geopolitics (as the news media popularize the subject), and academic geopolitics. Events since the 1990s have challenged the modern geopolitical imagination that arose to prominence in Europe beginning in the sixteenth century and its completely state-centered view of world politics. While state-centrism is still largely the province of both think-tank and popular geopolitics, within academic geopolitics the operation of such a state-based understanding of politics irrespective of geographical scale and other actors seems increasingly limited and limiting. Therefore, the conventional account of the spatiality (spatial organization) of world politics as totally territorial has been increasingly questioned, as it relies on three assumptions that have become problematic: (1) that states have exclusive sovereign power over discrete territories, (2) that the domestic and the foreign are separate and distinct realms, and (3) that the boundaries of the state define the boundaries of "society" (Agnew 1994). These assumptions have always been contestable, but with the end of the Cold War they came to be seriously undermined by changes in the practice of world politics. The continuing focus on territorial states has long overlooked the existence of two key aspects of spatial organization that intuitively relate to human experience and have been the subject of political geography for some time: spatial interaction and place making.

Spatial interaction takes place at any scale in the form of movement through networks. Developments in transportation and communication have allowed the stretching of relations of contiguity and distance, producing transnational networks. The very process of globalization operates on this basis. Even if the modern evolution of these technologies dates back to the nineteenth century (Hugill 1999), it is only since World War II that technological innovation in these and other areas has proceeded very rapidly, with the spread of mass jet travel and containerization, information processing and analysis (personal computers, business software, e-commerce, etc.), and telecommunications (the Internet, satellite television, and wireless telephones), some of it the outcome of Cold War military research. The explosion of movement that the world has been facing in the last sixty years would not have been possible without these technologies of time-space compression. The rapid circulation of people, information, capital, and goods around the world had already increased exponentially in the 1970s and 1980s, but it boomed in the 1990s when the end of the bipolar Cold War order met with the digital revolution.

Indeed, as noted previously, globalization had its ideological roots in the U.S. government's sponsorship of a "free world" economy during the Cold War. In the post–Cold War era, it came to continuously challenge the existing forms of political organization at different scales—for example, in terms of territorial sovereignty and political governance at the global, national, or urban and regional scales. The emergence of a "world of flows" based on transnational networks has therefore been undermining the dominance of the world of territories that characterized the Cold War geopolitical order. Already by the 1990s, the geopolitical nostrums of the Cold War, from military containment strategies to the imagining of fixed territorial opponents and command-and-control systems, became moot. Constructed much more in terms of nodes and networks than in territorial blocs, transnational networks of spatial interaction openly challenge the ability of states to channel and limit transactions across and within their borders.

Accounting for the spatiality of world politics entirely in terms of territorial states has also been challenged by a second trend that is intrinsic to the human experience of space and that also had been previously limited by the Cold War accent on the nation-state: that of place making. Places are natural locations that have acquired a social and psychological significance insofar as they ground political outlooks and projects in the settings of everyday life. Places can become grouped into territories during the course of state formation, and they can also be networked over space (Agnew 2009, 35). As we saw in chapter 4, identities are often expressed through their association with places and place-making activities. An explosion of ethnic, religious, and cultural identity conflicts has accompanied the end of the Cold War geopolitical order on a scale previously unknown, and all involve efforts at reshaping places to meet the political agendas of the groups in question. Some are the result of mismatches between ethnic and state boundaries that in some cases have had a long history (as in Kashmir, Spain, Ireland, Russian Chechnya, or Chinese East Turkestan); others reflect the lack of a basis in politics after communism other than national or ethnic differences (as in Yugoslavia). Finally, some came about as a reaction to globalization challenging traditional identities, especially in the form of religious radicalisms, as in Islamic jihadism in the Middle East and Protestant fundamentalism in the United States.

With the end of the Cold War in Eastern Europe, all those identities that had previously been repressed or marginalized by the overarching conflict produced an outburst of social and political protests. A case could be made regarding how much this actually contributed to accelerating the collapse of the Soviet Union beginning from the outer ring of its sphere of influence. Insofar as the world remains characterized by territorial states, it is not surprising that these challenged identities, formerly limited in their place-making activities by the geopolitical status quo, started to emerge as a power vacuum left scope for their articulation. In taking a nationalist stance, their territorial projection provoked a number of identity-based conflicts and a landslide of geopolitical change. Especially in those cases where ethnicities had historically been mixed in the same territory and a foundational sense of place was reestablished, the quest for national independence of one community clashed with that of a neighboring one (as in Bosnia and Kosovo). Paradoxically, as soon as these Eastern European identities succeeded in reaching independence and acquiring their own national states, they submitted their applications to join the European Union, implicitly recognizing the need for a supranational scale of political regulation in a globalizing world.

This is not to say that nationalism is the only possible outcome of place-making activities. As far as identity processes are deeply rooted in a sense of place, wherever human experience and its understanding overcome local horizons and recognize the similarity of the human condition, place-making activities could even scale up into a quest for a shared cosmopolitan identity based on a sense of place where the whole earth is recognized as a home to the human species. Indeed, the development of environmental awareness over the past fifty years has often transcended local issues to take the shape of a concern for the global environment. This could be seen as a postnational expression of place making. Meanwhile, as populations of different ethnic, national, and religious origins become intermingled in the developed, industrial, and postindustrial parts of the world, they bring with them the hopes, fears, and hatred of the places they come from and remain in contact with some of those they left behind. Identities in this case become increasingly diasporic (scattered around the world) and multiple (see Sen 2006). As much as this could contribute to a better tolerance and understanding of differences, wherever the process takes place too rapidly or too dramatically it could also produce an opposite outcome, as is often seen with global migration flows. Migrants could refuse to integrate with the hosting community as much as the latter could develop a symmetrical intolerance toward the "non-autochthonous" (the outsider) and his iconography (as seen with the 2009 referendum that banned the construction of minarets in Switzerland), and that in some cases can reach xenophobic extremes as with some far-right European populist parties.

With respect to globalization and related modernization challenging traditional societies, a resentment at the spread of consumer values and American popular-culture kitsch (as in many Hollywood films and American TV shows) and U.S. foreign policy has taken different degrees of intensity, operating in a variety of ways within those cultures and identities that feel most threatened. As far as religion is a fundamental component of many social identities, religious radicalism can appear as a self-defense reaction. The most evident of these reactions since the 1990s has been an outburst of Islamist radicalism in a hardened version of a "Jihad vs. McWorld" opposition (Barber 1994) aimed at American culture, values, and policies, especially those supporting countries unpopular with their neighbors, such as Israel. With the 2000s, the increasing use of suicide bombers in Islamic terrorism became its most extreme expression. But there is also unease at home in the United States with much of what goes for American and Western popular culture. This too is having political effects in the growth of populist groups oriented toward reestablishing some "older" image of American society. It is important to note, then, that even if based on a radicalized opposition to cultural or religious differences per se, as in the case of Islamic immigrants, the psychological mechanism can operate more generally. Beyond the specific case of anti-American jihadism, it could be seen in the destruction of the Buddhas of Bamiyan by the Taliban (2001), in the killing of Israeli leader Yitzhak Rabin (1995) by a Jewish fundamentalist (interrupting the Israeli-Palestinian Oslo peace process), and in President G. W. Bush's Manichaean view of the world as a field of struggle between Good and Evil (plus the role played by American Christian Evangelicals in support of this doctrine).

As each religious faith developed an attachment to specific sacred places that historically acquire a symbolic meaning for all believers, self-defense reactions imply specific territorial projections. In the case of nation-states, they can take the form of "cartographic anxiety" (Krishna 1994) toward borders when they are seen as defining national identities, as discussed in chapter 4. More generally, as the two-time political refugee Jean Gottmann knew too well, self-defense reactions can reach the extreme of dictating the complete closure of borders and the expulsion (or erasure) from territory of all the bearers of different identities or creeds. This can be seen as the goal of that religious radicalism expounding the expulsion of all the "infidels" from the lands of the Muslims to reestablish an Islamic caliphate, targeting even those coreligionists seen as deviating, to the point that among Arab Muslims, bin Laden's followers are commonly called *takfiri*, "those who define other Muslims as unbelievers." It is likewise the basis to the desire of those European nationalists and American nativists obsessed by the cultural dangers posed by Muslim immigrants (or foreigners in general) to expel immigrants, which has grown by leaps and bounds since 2001.

Since the end of the Cold War, these dynamics have put into question the traditional understanding of the spatial organization of world politics as based entirely on discrete bounded territories. Still, territoriality and sovereignty as such have not been eclipsed. It is more that the territorial state, itself a creation of Europeans since the sixteenth century, provides less and less of the monopoly over the means of violence, control over transactions, and basis to political identities that it once did, at least in many parts of the world. Theorists of speed and cyberspace or of "world flatness," however, miss the fact that events still unfold somewhere and in a world still divided politically into territorial units of one sort or another, which on top is riven by ethnic, cultural, and religious divisions reinforced by the self-defense reactions of identities feeling under threat. Rather than write in terms of flows *versus* territories and speed *overcoming* space, therefore, it is more appropriate to say that the regime of stable territorial sovereignties and relatively fixed spheres of influence associated with the Cold War has shifted toward a mixed territorial regime of flows between nodes in networks and territorial regulation and identity construction at and between a range of geographical scales—national, local, regional, and world-regional.

The coexistence of different spatialities of power and their interaction at different scales have therefore produced increasing geographical complexity, which has become more difficult to understand and manage. Since the 1990s various forms of resistance to aspects of globalization have emerged in a number of places. The explosion of Islamic radicalism can thus be analyzed in terms of the interplay between the new spatial interaction brought on by globalization and specific forms of place making associated with reviving traditional notions of social life. As Islamist terrorism took the form of transnational networks in the 1990s by taking advantage of globalized telecommunication networks, the U.S. "global war on terror" and its obsession with border security measures has attempted to reterritorialize the threat. But the flow of history seems to undermine such attempts, as very recent developments show. The global financial crisis that emerged in 2007 and the social and political democratic movements that swept through the states in the southern Mediterranean and the Middle East in the winter of 2010–2011 suggest that different forms of spatial organization are evolving in a world of seven billion people that we still do not really adequately understand, while climate change and disasters such as the meltdown of the Fukushima nuclear reactors in Japan in March 2011 question more than ever the absence of global governance appropriate to the problems humanity faces.

This complex global setting, therefore, defines the geopolitical context in which political geography will be made in the foreseeable future. It is very different from the contexts in which political geography was first "invented" and later remade. But it is not all new either. Even as "borderless risks" seem to proliferate, many features of politics remain relatively unchanged, as well as key features of human psychology and group identity. We shall now examine some of the topics that are of emerging significance in political geography. How these are dealt with is the likely task for making political geography over again in the early part of its second century. Of course, the process of making a field is an inherently uncertain one. As the geopolitical context changes, so will what we do and how we do it. What seems clear, and why this section on the geopolitical context has been so extended and complex, is that in the post–Cold War world the challenges facing political geographers have only become greater than they were in the past.

A MOBILE HORIZON

Trying to identify what is on the "horizon" of political geography requires divining from recent trends those likely to emerge into prominence because of their fit with the emerging geopolitical context. The purpose here is certainly not to exclude other topics from any future political geography or to offer an encyclopedic inventory of all contemporary trends in subject matter. In the geopolitical context just discussed, though, drawing attention to the current "making" of political geography allows us to show how active the redefinition and reworking of the field actually is. This is not to simply endorse what we consider the avant-garde and disregard what has been emphasized in the past. The purpose is to illustrate the flow of influence between the contemporary geopolitical context, on the one hand, and the making of political geography, on the other.

By and large, the new themes are not simply additions to the list identified in chapter 4—geopolitics, places and the politics of identities, etc.—but also raise metatheoretical and methodological issues (what concepts to use, how to conduct research, etc.) for established themes and theoretical perspectives as well. This is particularly the case with two of the three new themes explored in this chapter: geographical scale and normative political geography. But it is also the case with the first—the political geography of the environment—if only, for example, because climate change raises important questions for such established themes as geopolitics and the spatiality of the states, and also for the geographies of social and political movements.

Two background conditions are important in understanding much of what follows. First, a conversation rather than a dialogue of the deaf has begun to emerge between advocates of the three major theoretical "waves." Some examples are given toward the end of chapter 4. This perhaps reflects the deradicalization and routinization of what were once seen as totally opposing perspectives as their proponents achieve professional recognition and higher intellectual status. But this trend is also possibly the result of changing times, in which geopolitical instability and uncertainty makes the theoretical certainty (even if about the certainty of uncertainty in some varieties of postmodernism) of the different positions increasingly untenable. While an increasing theoretical eclecticism would seem both more necessary and more likely, then as students educated more broadly than past generations emerge into prominence to deal with an increasingly volatile and geographically complex world, in some cases post–Cold War geopolitical instability may produce an opposing trend: a hardening of perspective boundaries to defend against heresy.

Second, this world to which political geography directly addresses itself seems riven by identity-based (religious, national, local, or economic) divisions that take very different geographical forms from the state-based colonial and worldwide ideological systems that characterized the periods of interimperial rivalry (1875–1945) and the Cold War (1945–1989) during which political geography was previously made and remade. On the one hand, the world economy and many political movements (including religious ones) are organized in terms of spatial networks more than territorial blocs, as noted above. On the other hand, claims are made by leaders of dominant states and political movements on behalf of the welfare of huge geographical areas and their populations (such as the West, the Islamic world, etc.) as if these were homogeneous entities. These claims are often informed by a tremendous nostalgia for a period in the past (from medieval Islam to a golden age of state-based capitalism or the European welfare state) and for the "purity" of territories from "foreign" contamination, but now must confront a world in which people of diverse cultural origins increasingly live intermingled and interdependent with one another on a practical everyday basis.

In this context, three trends in political geography appear particularly significant. The first is a return to the old physical-human nexus of political geography, now in relation to how human threats to the natural "environment" are mediated by political institutions and movements. As climate change becomes increasingly connected to geopolitics, the risk of a reappearance of environmental determinism seems still far from being removed. So one task of political geography is to offer approaches that steer well away from this reinvention of what previously proved to be a dead end for the field. The second represents the continuing drift away from state-centrism toward according a central role for "geographical scale" in understanding the geography of power. If the global War on Terror has regressively renewed the importance of nation-states, the global financial crisis and the increased importance of the financial markets seem to subtract from conventional understandings of state territorial sovereignty. The third is the question of the nature of politics and its connection to the diversity of human experience. Political geography is finally addressing normative political issues concerning citizenship, democracy, group rights, and the role of intellectuals. These issues are increasingly fraught in a world of increased migration, economic interdependence, jurisdictional disputes over tax evasion, and the revitalization of ethnic and other identities. For each of the trends, a general introduction is followed by a number of examples to illustrate how its practitioners are presently making political geography often with clear reference to the current geopolitical context.

Political Geography of the Environment

The nineteenth century's holistic unity of physical and human/political geography was largely forgotten after World War II, since the spread of naturalized knowledge in political geography's historic canon (see chapter 3) had produced a major intellectual embarrassment. The political implications of environmental determinism in fact kept many geographers from developing new theoretical bridges with questions relating to the physical environment, leaving contemporary geography on the borders of "nature-culture" and "nature-society" debates. If the Sprouts were perhaps one of the few notable exceptions to the rule (see chapter 4), political geographers chose to work in different directions, abstaining from much dialogue beyond simple biological and physical analogies, as in the spatial-analytic perspective, or connecting to the environment solely on themes such as access to natural resources, as with the political-economic perspective.

The environmental movements of the 1960s and 1970s brought a renewal of disciplinary interest in the subject, often conjugated with its developmental implications in the political-economic perspective. By the end of the 1980s and early 1990s this had reached the postmodern/critical perspective (see chapter 4), especially with the growing interest in environmental security, dealing with topics such as how political theories relate to the environment, the politics of "green" movements and political parties, and the study of environmental policy making (Carter 2007).

Since the 2000s, a number of trends have reinforced the perception that a return of the environment on the horizon of political geography may gain increasing centrality.

Physical geography also grew separated from political geography; only recently have some of its practitioners come to realize how much latterly it was influenced by Cold War dynamics. In fact, from the second half of the 1970s onward a rapidly growing earth-science community began to receive increased funding from the United States government and its allies, as political leaders started to fear the potential impact of environmental problems of one sort or another on the Cold War order. Moreover, the access to technology originally developed under military programs, such as earth observation satellites and since the 1980s supercomputing and computer networks, boosted the methods available to the field. So an ever more dominant earth-science community increasingly absorbed many aspects of physical geography. As a result of this scientific hegemony, and not just in a marginal way fostered by technological advantage, climate modeling became increasingly important, to the point of indicating global policies to address the "problem" of anthropogenic climate warming, which became increasingly politicized.

Given that in the dominant discourse climate change is seen as a consequence of modernization affecting both nature and society (as well as in its central tenet of "sustainable development"), the call to cross disciplinary boundaries and engage directly with the epistemology of climate from a geographical perspective has come only recently from within the earth-science community (e.g., Hulme 2008, 2009a, 2009b, 2010; Bailey 2008). As the difficulties of integrating the social and human dimensions of climate change became more evident, an epistemology more inclusive of the cultural and political dimensions of climate was seen as an important contribution to filling the disciplinary gap. Touching a central debate in contemporary political geography, Hulme recognized that an advantage of such engagement for the climate-change community would come from geographers' ability "at understanding the subtleties of how knowledge, power and scale are inseparable" (Hulme 2008, quoting Cox 1997). Other calls have come also from within political geography, including the recent idea of "re-naturing political geography" (Cox, Low, and Robinson 2008, 183–262).

Another reason for a return of the environment to the horizon of political geography comes precisely from the general ignorance of the disputes and resolutions that took place in political geography on the subject of naturalized knowledge (chapter 3). With the continuing increase in complexity, monocausal explanations still seem to be quite tempting to many, opening up the risk of a renewal of environmental determinism or "neodeterminism" as it is termed by some, from the spread of neo–

environmental determinism (see Sluyter 2003; Merrett 2003; Judkins, Smith, and Keys 2008), to those varieties where physical conditions are said to directly cause differences in economic and political development (e.g., Landes 1998; Sachs 2000, 2005), to those, closer to mainstream ecology, such as the Worldwatch Institute, who draw a causal connection between a vast range of environmental disasters and "global consumerism" (L. Brown 1981, 1999). In none of these cases are the roles of political economy, cultural invention, or institutional arrangements accorded much more than marginal status in accounting for contemporary global environmental and development problems. It is as if the past one hundred years of political geography and all the debates over environmental determinism and the mediating roles of discourse and institutions had never taken place. Needless to say, a priority for political geography should lie in helping to correct this ignorance, avoiding those conceptual traps that have already been proven empirically wrong and politically dangerous.

While the politics-environment nexus may have been disregarded within Cold War political geography, we shall now look at some examples that show how political geography can fruitfully engage with different aspects of the environment, from the analysis of how the uneven geographical distribution of resources could contribute to conflicts to the politics of the environment and the environmental connection to development issues. Moreover, the increasingly pervasive topic of climate change—perhaps as pervasive today in the public arena as security discourse—has opened multiple connections to development, security, governmentality, scales of governance, and many other issues. We shall therefore consider some examples for each of these areas: resource wars, environmental security, and some of the many implications of climate security in their relation to state sovereignty, violent conflicts, and human security. The difference from the past is that as geographers and other social scientists increasingly claim the topic of the physical environment as their subject matter, the very idea of nature is now seen as being political at its foundation.

Geopolitics of Resource Wars

One environmental theme that obviously connects to political geography is that of resources and the degree to which they have or might figure in conflicts. Nonrenewable resources in great demand, such as oil and gas, are clear candidates for bringing about conflict, as in Kuwait (1990), Georgia (2008), or Libya (2011). Possible "flash points" for resource conflict, particularly in the Middle East and Central Asia, have become the major focus of U.S. government defense planning (Klare 2001). But whether food and water shortages can directly precipitate conflict is subject to greater doubt than is possible conflict over world oil supplies (Rogers 2000; Lowi and

Shaw 2000). Wolf and Hamner (2000, 123), for example, note "countries seem not to go to war over water." There is a danger of presuming that resource conflicts must invariably escalate to warfare.

Nevertheless, the links among environment, population, and military conflict are well worth investigating, not least to see how institutions mediate between physical geographical conditions and intergroup and interstate conflicts (Homer-Dixon and Blitt 1998; Lowi and Shaw 2000). In an increasingly globalized world, "risks" of all sorts, other than traditional military ones, are coming to dominate thinking about security and conflict (Albert 2000; Ó Tuathail 2000a). As natural resources are inevitably located at fixed positions, a long-established tradition has connected their uneven geographical distribution with geopolitics, incidentally offering an example of how the spatial-analysis and political-economic perspectives can actually come together. This connection has in fact long been the focus of economic and political geographers who value a (historical) materialistic approach, the classic example being that of the role of oil in geostrategy and geopolitics, from Hitler's attempt to occupy the Caucasus during World War II to the First Gulf War, the U.S. militarization of the Persian Gulf, and the renewed centrality of the "oil heartland" (Iseri 2009; Güney and Gökcan 2010).

As military security against "terrorism" became a political priority in the past decade, energy security also moved to the central stage of political concerns, and there have been a number of attempts at reading U.S. geopolitics in the Middle East as the main result of the global competition for oil, with a particular turn in the discourse after the beginning of the War on Terror (e.g., Le Billon and El Khatib 2004). As the 2011 war in Libya has clearly shown, this may not be solely a concern for the United States. Still, one thing is to weigh in each case the role of specific resources within geopolitical decisions in relation to other factors, including ideologies, alliances, political psychology, and risk perception, and another is to attribute to the competition for resources an absolute explanatory power, a move that inevitably falls under the label of neodeterminism.

At the intersection of political geography and resources, an excellent example of how the risk of neodeterminism can be avoided even when investigating the relationship between oil and conflict at the intranational scale is offered by a classic study on Nigeria, the world's thirteenth-largest oil producer and a country whose economic growth is almost entirely based on oil (M. J. Watts 2004). The efforts by a number of Niger Delta states to control the oil and its revenues and the struggle for self-determination of national minorities, in the context of a crisis of rule in the region, with growing insecurity and violence even within communities, and of a dispute between the federal state and the alliance of littoral states over offshore revenues, may easily be portrayed as a curse for Nigeria. Though, building on

Mamdani (2000), as oil is "a national resource on which citizenship claims could be constructed," and in consideration that "oil's contribution to war and authoritarianism builds upon pre-existing (pre-oil) political dynamics," Watts comes to the conclusion that oil has rather become "an idiom for doing politics as it is inserted into an already existing political landscape of forces, identities, and forms of power" (2004, 73, 75, 76).

Global Climate Change and Local Impacts

Another theme that connects the physical environment with political geography is that of global climate change and its many implications. Climate warming is affecting a wide range of natural systems worldwide. Relatively small variations in global mean atmospheric temperature affect the water cycle, though with a range of very different effects and an uneven geographical distribution. For example, while atmospheric warming produces a global increase in rainfall, in the Atlantic there seems to be an impact in terms of greater strength, frequency, and southern reach of hurricanes, while in the Sahel and the Mediterranean it equates to greater desertification. Not all warming is detrimental: at high latitudes, in Russia and Canada, an increase in their potentially cultivable land is expected. In the cryosphere, climate warming means ice melting. While there is evidence that the majority of world glaciers are retreating, in the case of those in the west and north-central Himalayas, there is also a concern that this could affect tens of million people who depend at least partially on melt water for their drinking supplies. The predictions of when such changes could happen remain largely questionable, but a protracted melting of ice and glaciers over the land would produce one of the most publicized and feared consequences: a rise in the mean sea level, which could threaten densely populated deltas, coastal settlements and small island states, affecting hundreds of millions.

Climate change has particularly boosted global perceptions of the political and economic consequences of sea ice melting in the Arctic Ocean. Projected increases in global mean temperature by the end of the century (strongly amplified in the Arctic) will reduce consistently the sea ice, as it has already been observed during the September minimum between 2007 and 2011. This has already catalyzed the attention of many political geographers on the Arctic, which is now seen as a new arena for resource mining and maritime shipping (e.g., Dodds 2008, 2010; Powell 2008, 2010; Dittmer et al. 2011). The increased accessibility of the Arctic Ocean and its implications have in fact started a geopolitical "scramble for the Arctic" among the five littoral states (Canada, Greenland, Norway, Russia, and the United States) that would directly benefit from extending their sovereignty over a greater area of the Arctic Ocean floor as a consequence of reduced sea ice. The historical attempt to reduce the dynamic Arctic environment to "a fixed division between land and water" is further complicated by the claims of indigenous northern people, as well as by the position of the EU, questioning the state-centered focus of development in the region (Gerhardt et al. 2010), which instead could present an opportunity to experiment with transnational governance (O. Young 2009).

Nevertheless, Russia was first in laying its claims to sovereignty when in 2007 an automated submersible planted a titanium Russian flag on the ocean floor beneath the North Pole. Canada quickly opposed this on the grounds that the twenty-first century is not the fifteenth century, when one could simply go around planting flags to claim sovereignty. To be sure, all the littoral states have promoted scientific missions to improve the topographic mapping of the seabed (until recently relatively uncharted due to ice cover). Russia in particular aims at connecting the Lomonosov Ridge to its Siberian continental shelf, a move that would extend its exclusive economic zone (EEZ) to encompass the North Pole (the EEZ is calculated to two hundred nautical miles from a baseline, but the topography of the latter in the Arctic is still controversial at a number of locations). Therefore, new seismic and bathymetric data could substantiate the conflicting claims, in order for the appropriate commission to decide on the EEZ extensions under the UN Convention on the Law of the Sea (UNCLOS).

In fact, the sea-ice reduction has already improved each country's accessibility to their current EEZs, and the multiple extension claims aim at maximizing prospects of exploitation of the yet undiscovered fossil fuels that may exist in the Arctic. Recent geological assessments suggest they could represent about 30 percent and about 13 percent of the world's undiscovered natural gas and oil, respectively. Based on the current EEZ extension, a recent study, using the new Arctic Transport Accessibility Model (ATAM), has calculated an increased maritime access by 2045–2059 that would benefit especially Greenland (+28 percent), Canada (+19 percent), Russia (+16 percent), and the United States (+5 percent). If the UNCLOS will award the maximum plausible extensions to the EEZs, the national figures would respectively rise to increases of 37 percent, 32 percent, 29 percent, and 11 percent (Stephenson, Smith, and Agnew 2011). In consideration of the current dependence on fossil fuels and their increased prices, such figures explain the enormous interest of the littoral states in extending their sovereignties over the Arctic, though this will certainly elevate the risk of environmental damage from spill and discharge.

The sea-ice reduction in the Arctic also means improvements to maritime transportation, as a number of routes are already opening up for a longer part of the year. The trans-Arctic routes have the potential for significant distance savings: 40 percent via the Rotterdam–Yokohama Northern Sea Route (compared to the Suez Canal), and 33 percent via the St. Johns–

Yokohama Northwest Passage (compared to the Panama Canal). It is predicted that with the considerable decrease in sea ice, Canada, Greenland, Russia, and the United States will enjoy nearly year-round maritime access to their respective EEZs (using ships with light icebreaking capability) by 2045–2059 (also allowing faster travel speeds within regions already accessible today). Thus another sovereignty dispute is that of the Canadian government, which claims the Northwest Passage as within Canada's internal waters, with the United States, which claims the passage is an international strait. The dispute has not been resolved yet, but if Canada's claim is recognized, it will have the right to restrict transit by other states' vessels (Gerhardt et al. 2010).

Meanwhile, it has been discovered that the impact of sea-ice melt on Arctic transportation is not single but double (Stephenson, Smith, and Agnew 2011). All Arctic states will face declining possibilities for inland "winter roads," those temporary ice highways indispensable for remote communities located far from navigable inland waterways or coasts. This would also impact mining, energy, and timber industries in these areas. In underscoring the acute biophysical sensitivity of the Arctic to climate change, this study also highlights the dangers of simplified characterizations of social responses to climate change and the importance of increased collaborative and interdisciplinary work between earth scientists and political geographers (see figures 5.2 and 5.3).

From Environmental to Climate Security

Modern geopolitical thinking is dominated by a view of security that is essentially territorial: friendly and enemy blocs of space are posed against one another as spheres to be protected or contained against military and economic threats. While the security situation faced by states and the world's population is now both more global and more diffuse (Ó Tuathail 2000a), as a whole set of problems are planetary in scope (from nuclear accidents to global warming, from solar magnetic storms to global pandemics), yet they must be addressed by the existing territorial system of government, which despite the huge priority given to security in the 2000s, may not be ready to face such emergencies, as a long list of examples could illustrate, including the 2004 Indian Ocean tsunami, the 2005 Hurricane Katrina, the 2010 eruptions of the Eyjafjallajökul volcano in Iceland disrupting international flights, the Deepwater Horizon major oil spill in the Gulf of Mexico, and so on. At the same time, environmental problems afflict specific communities in particular places (technological and environmental hazards, air and water pollution, etc.), which could pit social groups and political movements against one another in the struggle to avoid and displace environmental dangers.

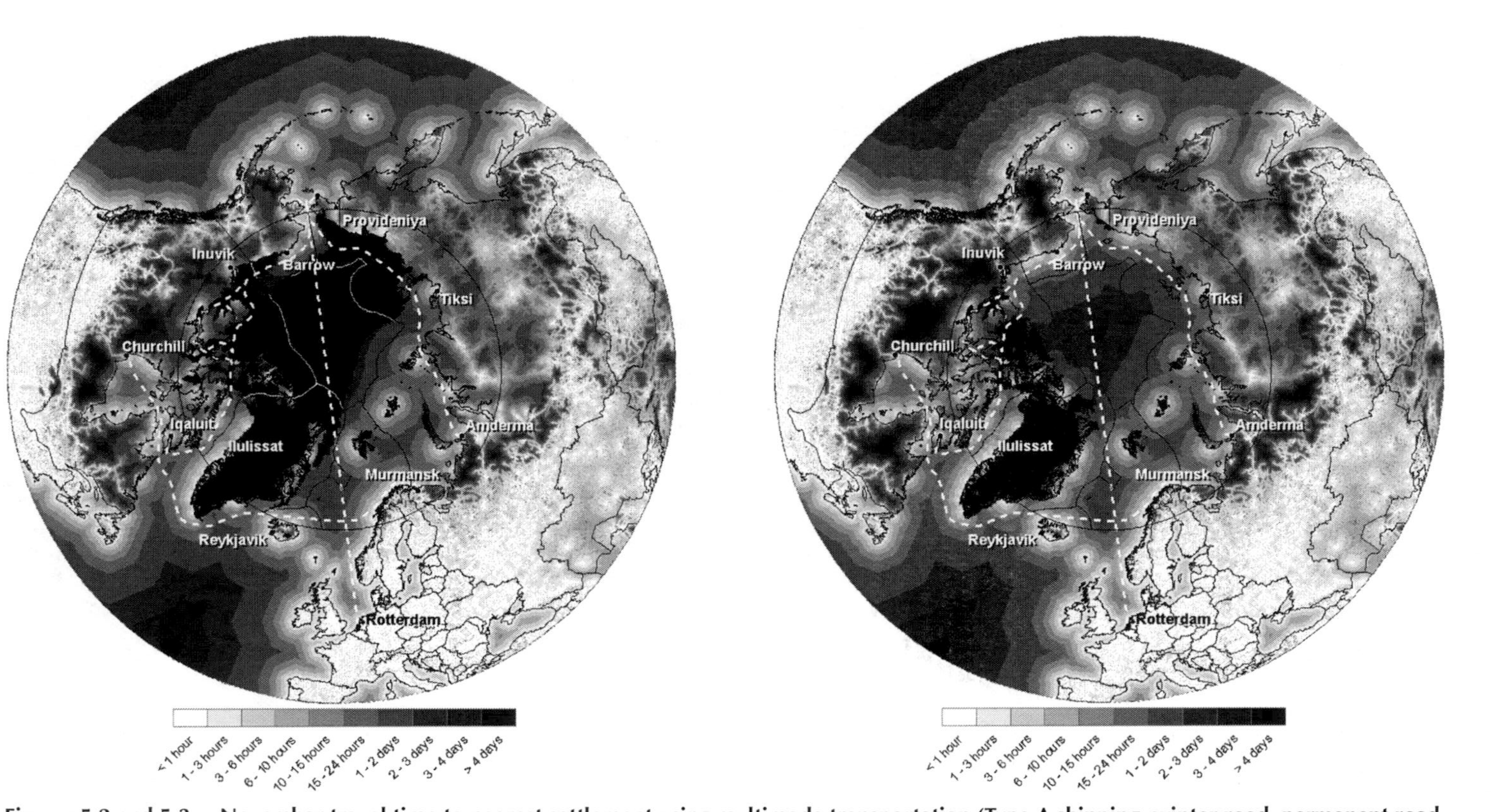

Figures 5.2 and 5.3. November travel time to nearest settlement using multimode transportation (Type A shipping, winter road, permanent road, rail, walking) for baseline (2000–2014, left) and midcentury (2045–2059, right). To isolate climate change impacts, all human settlements and permanent transportation infrastructure are assumed unchanging between the two time periods. By midcentury, dark tones in continental interiors reflect longer travel time to nearest settlement (TTNS) owing to reduced winter road potential. In oceans, lighter tones reflect shorter shipping TTNS owing to reduced sea ice. Areas still inaccessible by midcentury are shown in black. The dotted lines represent the trans-Arctic routes.

Source: Redrawn from Stephenson, Smith, and Agnew (2011)

In each case, however, threats can no longer be spatialized only in national terms, as if state boundaries neatly matched the negative externalities of environmental danger. From this point of view, security is increasingly globalized and localized. A number of scholars have proposed studying environmental security in terms of the "blowback" and "boomerang" effects that arise in modern global society in which the traditional calculus of "risk" has changed dramatically (Ó Tuathail 2000a; Albert 2000; Dalby 2002). In this understanding, people are increasingly conscious of the side effects of economic growth and weapons production and thus begin to call the traditional rationality of cause and effect into question. Consequently, the environmental side effects of other human actions have become more important politically. The German sociologist Ulrich Beck (1992) has coined the phrase "risk society" to describe this new modernity in which the unintended consequences of industrialization and consumption are acknowledged and confronted.

This emphasis has encouraged a rethinking of conventional views about security threats and political responses to them. One radical conclusion is that insecurity always may have been more characteristic of human societies than is often acknowledged, especially considering the literature on the perception of risks (e.g., Schwarz and Thompson 1990). Industrialization and the insulation of a large number of people in developed countries from visible dependence on the natural world may have encouraged a certain hubris about the security of the human condition. With the rise of "risk society" the sense of human vulnerability has returned. In particular, environmental consciousness has coincided with the moral critique of high mass consumption in some places when many people are destitute elsewhere, suggesting a link between environmental degradation and social justice; global pollution, pressure from farmers, and commercial lumbering are creating serious ecological "hot spots" around the world where the survival of many plant and animal species is in serious question; and military security has increasingly produced environmental insecurity (witness the possible role of U.S. weapons labs in the 2001 anthrax scare, worries about disposing of nuclear and biochemical weapons stocks, or the health side effects of wars) and the "securitization" of environmental issues and especially climate.

In fact, the social, political and economic consequences of climate disruptions have long been a concern of the intelligence and defense communities in the United States and elsewhere. Since the 2000s, while security became a national priority in the context of the War on Terror, it also came to incorporate within its realm of concerns climate warming. In 2007 the topic was even included among security issues to be addressed within the UN Security Council. Therefore, militaries have started to carefully consider the impacts that climate changes could trigger on local communities, especially in terms of potential violence and conflicts that would escalate to wars, as well as the related risk of climate-induced migrations.

While it is possible that a genuine rise in environmental awareness may have finally reached at least some members of the military, which among its tasks includes relief efforts to disaster stricken areas and populations, nonetheless warfare activities as a whole are responsible for a number of environmental problems. It is estimated that they contribute about 10 percent of global carbon emissions; major environmental damage resulting from war-preparation activities have been documented at a number of sites—for example, the island of Vieques, Puerto Rico, used for over sixty years as a shooting range—not to mention immediate and indirect wartime impacts on landscape and the environment. These studies fall in the area of so-called warfare ecology and are seen as having policy implications, including military policy making and planning (as the recent strategies to "green" U.S. army operations show), and could also contribute to the transition of former military sites to conservation purposes (see, e.g., Machlis and Hanson 2008).

Nevertheless, the so-called securitization of environmental issues, and of the climate debate in particular, may have been resented by many as a sort of pitch invasion, especially considering that the environment has long been one of the main grounds of moral critique of modernization. Therefore, it is not surprising that the appropriation of the environment as a security issue has met with the criticism of social scientists and of some political geographers in particular. An immediate response has been that of connecting climate security to biopolitics (e.g., Dalby 2011) or of exposing the governmental regulation of human populations and the securitization move as functional to the dominant discourse of fighting against the "natural" anarchy of the world (e.g., Barnett 2009). While a number of studies have been investigating the impact of climate issues on violent conflicts, this specific link has been opposed as speculative and lacking evidence, especially considering the high variability of social responses to change. Therefore, its presumed causal relation has been found to fall into the ever-expanding range of neodeterminism.

However, according to Nordas and Gleditsch (2007), this area could be investigated by political geographers in a number of ways: through coupling climate models with conflict models in order to predict where climate-related conflict hot spots may develop; by considering what kind of violence could result from climate change (such as in Rwanda and Darfur); by balancing the positive and negative effects of climate change and the related strategies of adaptation; by disaggregating the effects of climate change in conflict models both in terms of geographical variations and types of change (total rainfall increases, but some areas will become drier); and by focusing on the consequences not just for rich countries but also for the poor ones (which are likely to be more affected)—though there may well be a serious risk of determinism in placing too much trust in the use of conflict models. Analyzing the link between climate and migrations first exposed by Norman Myers (1993), some have realized that this may in fact be triggered by more complex causes, including food, water, and health, among others, rather than climate change per se (Eakin 2006). While at least in the case of small island states, such as the Maldives, climate change impacts may not necessarily produce climate refugees, as the inhabitants of these islands ask for policies that would allow them to remain where they are even in the event of a sea-level rise. In the face of attempts at squeezing the new problems of climate security into older conceptions of national environmental security, Liverman (2009) proposes linking climate change to the human security agenda instead of connecting it to the national security one, as with themes such as water and food security.

Water Security

Another scarce resource frequently considered to entail geopolitical conflict is water, to the point that access to potable water is often said to be the defining world crisis of the twenty-first century (e.g., Chartres and Varma 2010). While a number of examples from around the world—from California and the larger region served by the Colorado River to the American Great Lakes region, from the Middle East to China, from the Himalaya region to Africa—confirm that access to water is a major problem for many, the simple connection usually drawn to local water availability and to the geographically differential effects of climate change hide the fact that the world water problem is predominantly a political one (Agnew 2010).

By comparing water availability with water consumption, the three most populated countries—China, India, and the United States—stand out as the largest water consumers. In relative terms, however, water use per person in the U.S. is about triple that in either China or India. In Central Africa, endowed with relatively good water resources, consumption is only about 2 percent of that of the average American. The per capita differences largely depend on relative population sizes, differences in capacity to tap ground water and distribute it, and overall rate of use relative to availability. The typical generalizations drawn from comparing water availability with consumption is that the world's population distribution doesn't match that of the water resources, that there are dramatic differences in ability to exploit available water resources, and thus that there is a significant human water deficit on a world scale (see figure 5.4).

Still, the apocalyptic view according to which water is inevitably going to trigger conflicts all over the world is deterministic (Agnew 2010), as in those studies following the Wittfogel-inspired approach that considers

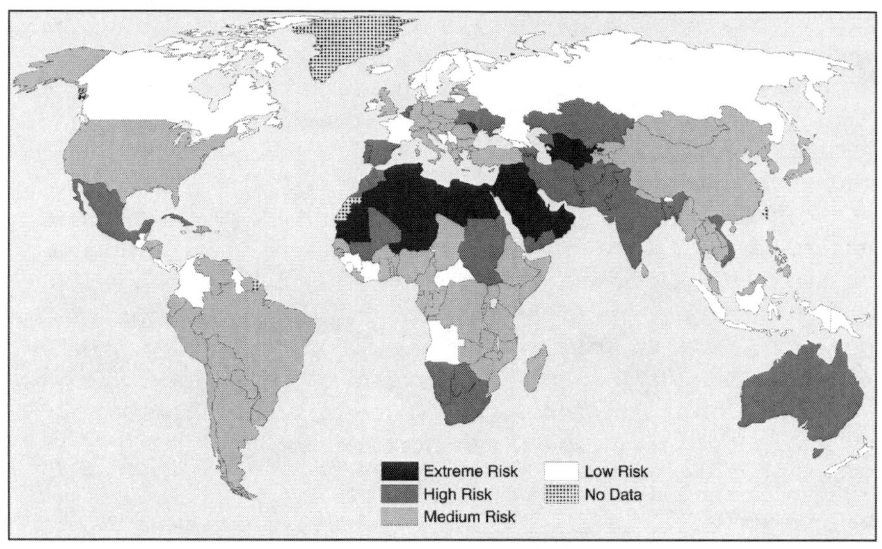

Figure 5.4. Water security in the world. By comparing water availability to water consumption it is possible to estimate the water security risk for each country.
Source: Multiple sources

water as mandating centralized or despotic control over waterworks (e.g., Solomon 2010), or in the accounts that see water as an independent force in geopolitical conflict (e.g., Pomeranz 2009, on the Great Himalayan watershed between India and China), or in those narratives were water is seen as part of the larger drama of economic and urban growth versus nature, an instrumental ally of elites versus its supposed natural state of being. In all three narratives, water is accorded an independent causal power that eliminates politics in any sense of deliberation, decision making, negotiation, and accountability. Yet it is well known that water conflicts rarely if ever escalate into warfare, if only because water itself can be shared and cooperative governance can allow for win-win rather than zero-sum outcomes.

Global Climate Governance

Since the 1985 UN meeting on climate change through a number of international summits, the dominant policy direction offered by the earth-science community has mainly been focused on climate-change mitigation through reducing the volume of human-made gases entering the atmosphere. The Kyoto Protocol (1997) aimed at reducing CO_2 emissions was based not only on each country's different contribution but also on its stage of economic development, and was planned to develop complex schemes of carbon trading among states. Because of the link between CO_2 emissions and fossil-fuel energy, the United States, an advanced economy heavily culpable in terms of contribution to total carbon emissions over the years, has not ratified the protocol, therefore contributing to its failure. This decision was defended through a denial strategy, shared also by OPEC member countries, based on the idea of "questioning the science and highlighting the risks of responses" (Barnett 2009).

Meanwhile, China seems to have recently overtaken the ugly primacy of the U.S. as the largest emitter of fossil-fuel CO_2, and it has been calculated that "developing economies as a group have greater emissions than developed economies" (Boden, Marland, and Andres 2010). This has led some to question the effectiveness of emissions mitigation in developed economies without the commitment of China and India. In response, Waldoff and Fawcett (2011) have shown that this is no reason to delay actions already agreed to at the July 2009 Major Economic Forum by the G8 leaders; even if developing countries delay their mitigation efforts to 2025 or 2050, the contribution of developed economies to reducing emissions remains essential (although it would obviously be better if all countries began such efforts immediately). Still, at the Copenhagen Conference (2009) a shift among centers of geopolitical power was evident: rather than the EU, the Anglophone countries, and the OECD member states, the decisions turned out to be mostly in the hands of China, India, Brazil, and South Africa. The Copenhagen agreement recognized that global temperature should not increase above 2 degrees Celsius, but the conference outcome was generally perceived as a failure in terms of its ability to produce any more stringent action by the participating states.

Why is reaching agreement on a global emission reduction policy so difficult? Is it a problem of risk perception, or is it related to the shift between centers of geopolitical power? In an interesting article on why people misunderstand climate change, Chan Ziang (2011) argues that a limited mental model is often used to analyze climate change. This lets people mistakenly believe that it is possible to stabilize CO_2 by keeping emissions at current rates. Drawing from findings in cognitive and developmental psychology, Ziang explains this misunderstanding as arising from an error in people's ontological assumptions: the problem is associating climate with a static model (useful in understanding objects), while its ontology is that of a dynamic process with unique features in terms of how it develops over time.

While this may provide an account of the general problem of misperception, including even that of some decision makers or principal actors, Phyllis Mofson (1999) instead would probably respond to the above by observing that the problem is rather more inherent to the relationship between the complexity of the environment, on the one hand, and the hierarchical governance structure that is addressing it, on the other. Considering her notion of "ecopolitical hierarchy," developed to represent the scale and attributes of environmental problems (figure 5.5), if ecosystem complexity

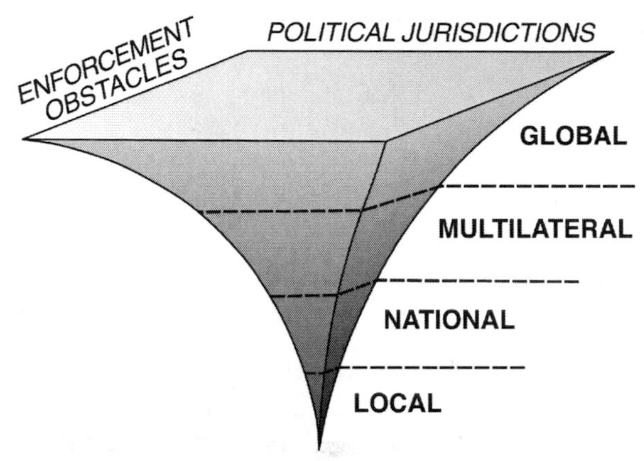

Figure 5.5. The ecopolitical hierarchy.
Source: Mofson (1999, 247)

increases as one moves up the geographical hierarchy, so do the number of actors, political jurisdictions, and barriers to resolving any particular problem. As a problem is associated with higher levels in the hierarchy, so the possibility of resolution becomes more difficult. As a result, global climate change is much more difficult to manage than would be some local pollution problem. This does not mean that local or regional environmental problems are not also exceedingly complex and difficult to manage (R. W. Williams 1999). Nevertheless, even before Copenhagen it was clear that the very approach to climate change and its current (lack of) governance could be questioned way beyond the usual discourse of climate-change deniers.

Facing the evident lack of anything that could even remotely resemble the attributes of earth-system governance, Hulme (2009b) recalls how the social meaning of climate has changed historically and geographically. As initially a Cold War–era construction according to which humans can change the physical properties of climate, a hegemonic earth-science community has promoted the disputable idea that future climate is predictable through better data and more powerful modeling, and therefore that global climate stabilization is actually possible. Hulme and others have questioned this approach that fetishizes carbon, focused as it is on mitigation policies, and have suggested instead the need to concentrate more on adaptation policies, acting through different scales of governance, above and below the nation-state, through a number of smaller steps in a plural and polycentric way, including tackling climate-warming gases other than CO_2. According to Hulme, not a minor part of the problem of over twenty-five years of attempts at reducing CO_2 seems to be the universalistic approach to climate warming it presupposes, culturally stemming directly from the Cold War. The dominant narrative has been based on the threat of climate chaos, to avoid which, since the 1990s, earth scientists have promoted an authoritarian solution based on a centralized global governance (the UN) that would bind in a global treaty almost two hundred states to commit to reducing the hardest part of their emissions, those from fossil carbon energy (which account for half of the problem). Still, "the world has no meta-narrative—whether offered by science, ideology or religion—that can provide a universally accepted foundation upon which a centralized global governance regime can be built" (Hulme 2010, 18).

The evidence for an overall increase in global atmospheric temperature is not questioned here, but Hulme's radical approach interrogates the reductionist move of shrinking the future of the earth to a single universal problem (climate change) that should be targeted with a single universal solution (reducing carbon emissions) through a global top-down mechanism enforced by a single centralized authority. While there is scientific consensus on climate warming, the future may not be reduced to this singular climate discourse, even if computer models' simulations keep getting increasingly sophisticated. To this purpose, it may be useful to get some historical perspective. Hart and Victor (1993) have reconstructed how the issue of climate change came to the fore during the Cold War, with all of the negative baggage that usually comes with it. After the UN Conference on Human Environment (Stockholm, 1972), the greenhouse effect started to be seen as a problem with potential geopolitical implications. This contributed to funding for climate research, allowing the current theories of anthropogenic climate change to develop. If a link between nuclear weapons and climate change as "comparable threats to civilization" was first established by von Neumann (1955), a rise in environmental awareness could be conceptually related to the disquieting knowledge that the invention of nuclear fission reversed the historical order that saw man as inevitably subject to nature, inspiring the search for a similar pattern of overcoming nature as that in the Industrial Revolution. Technology could still conquer all, including the ability to reverse engineer what had already been brought to pass through environmental damage. Beyond some apocalyptic sentiments, the rise of environmental movements may well be also the result of the possibility of political action it offers to the individual and to a community (not just in terms of change in individual behavior but also because corporations and politicians have realized this is a sensible issue), as opposed to the impossibility of political action toward the Cold War nuclear proliferation that had piled up enough power to destroy the entire planet several times over while no one was able to stop it. Therefore, much of the dominant contemporary discourse about climate change and its geographical consequences has roots in the hubris about centralized, top-down engineered solutions to environmental and military problems characteristic of the Cold War era.

Nuclear Security

While the military nuclear-political question has produced since its origins in the 1940s a debate with a number of different positions, mostly bound by the context of the Cold War (Zuckerman 1982; Deudney 2007), the nuclear debate has after the long crisis that followed the Chernobyl disaster (1986) focused more on nuclear energy, based on the primary observation that, being almost carbon free, it could be a cheap alternative to fossil fuels. This claim was always problematic given that the nuclear industry has long been heavily government subsidized in many countries and the costs of waste disposal and accidents rarely figure in comparative cost scenarios. Obviously, such an economics has also been thrown into question by the as-yet-unknown implications of the 2011 multiple meltdown in three reactors and spent fuel pools at the Fukushima power plant in Japan. Even if available data has not been released, this has clearly been the worst technological accident ever, with consequences well beyond Japan, considering that it was the largest accidental release of radionuclides into ocean waters ever, which by affecting marine ecosystems could then enter the maritime food chain. This has already reshaped nuclear agendas and policies in Germany, Switzerland, and Italy and may well affect the future of the entire nuclear industry. Certainly, as soon as more data becomes available, political geographers will start to work on its many implications, including the most neglected question: the political geography of nuclear waste disposal (Macfarlane 2011). In the case of plutonium spent fuel rods, radioactivity declines to levels considered safe only after twenty-four thousand years. Considering that recorded human history goes back about six thousand years, Schell (2011) suggests that we pause and think for at least the next twelve thousand years. No one can as yet guarantee a perfectly sealed radioactive waste container in a safe geological area for such a time span.

One hard lesson we certainly need to learn from Fukushima is that human perception of risk is far from perfect. Nuclear plants are designed on a safety analysis based on the concept of resisting "credible events." In the Fukushima case, the earthquake magnitude (M9) and the wave height of the related tsunami (fourteen meters) exceeded the credible event on which the plant's design was based. Unfortunately, the probabilistic risk assessment methodology employed had inherent difficulties in incorporating all the geologic data that earth scientists had made available and at interpreting it in a wider context. Moreover, the plant operator ignored two warnings shortly before the earthquake from geologists about the dangers facing the plant. Currently, thirty-one countries operate 437 nuclear plants, and sixty-five more countries have expressed interest in starting nuclear power for energy production (IAEA 2010). Obviously, the official reason for their interest is electricity generation that is free from carbon emissions. But since the ground to cover between a power plant for civilian use and one to produce nuclear weapons is not very great, once a power plant is in place, it is reasonable that a number of the states having such plants may actually try to develop nuclear weapons for use in developing geopolitical deterrence capabilities, as has long been feared about Iran's civilian plans for uranium enrichment—and as the recent agreement to develop nuclear energy between Saudi Arabia, sitting on one of the largest fossil fuel repositories in the world, and the U.S., again under the claim of producing electricity, clearly shows.

As we have briefly tried to show, there is certainly much to do before political geography could claim back the physical environment within its epistemological compass, and in so many directions, especially considering the politicization of climate change in contrast to the current depoliticization of the public sphere in which it must be debated (Swyngedouw 2010). Discussing nuclear and water security could be at the heart of an environmentally informed political geography. More generally, to better understand how concerns for nature remain at the epistemological foundation of the field, it may be useful to recall the overall importance of cultural and political ecology. Bruno Latour (2004) has shown that the very idea of nature in the West has been a historically situated social representation, and is thus political at its very heart: "Nature is not a particular sphere of reality, but the result of a political division . . . that separates what is objective and indisputable, from what is subjective and disputable," thus giving us the Western distinction between an idea of nature (viewed as objective) and that of a society (seen as subjective). Consequently, "we cannot characterize political ecology by way of a crisis of nature, but by way of a crisis of objectivity" (2004, 231, 22).

GEOGRAPHICAL SCALE

In political geography, geographical scale is usually thought of in terms of either the fixed or the emerging dominance of one level over others in political organization and behavior. In particular, the national and the global have achieved a privileged status as the geographical scales at which politics is said to be determined. In the former case, this is because a spatial ontology based on states rules out autonomous sources of political agency at other scales. In the latter case, a spatial ontology emphasizing global relationships, either geopolitical or economic, rules out much if any autonomy at other scales, except that resulting from a state's or a region's relative location in global geopolitics or within the global division of labor. Each is also seen in either/or terms. Each excludes the other as causally significant in its own right, except as a moment in the workings of the other. Needless to say, other geographical scales receive short shrift indeed.

These orthodox understandings of geographical scale have been called into question by recent trends in the workings of world politics, world economy, and the global environment, particularly the emergence of newly powerful global city networks (e.g., Beaverstock, Smith, and Taylor 2000), global political networks (of transnational companies, international migrants, drug dealers, terrorist groups, etc.) (e.g., Held 2000), cross-border regional compacts (e.g., Blatter 2001), and political movements concerned with large-scale issues relating to the physical environment, globalization, human rights, and arms control (e.g., Princen and Finger 1994; Price 1999). All of these enterprises and networks operate across scales rather than restricting their activities to only one. With the value of hindsight, the past also now seems less singularly national or global than it once appeared (for a brilliant account to this effect, see Wolf 1982). For a long time political geographers remained intellectual prisoners of the first age of geopolitics in the 1890s. The centrality of the nation-state in political geography left the dominance of the national scale unquestioned for a long time. In the last decades, and especially with the emergence of a global scale and the increasing attention to local scales such as the regional and the urban, political geographers have begun to see the past in terms of the workings of processes across scales. This seems likely to become an important feature of a non-state-obsessed but equally non-globe-entranced political geography (see Marston 2000). In highlighting the historically shifting relationship between scale and sovereignty (Claval 2006), we see that it connects also with how governance is organized, as with the issue of climate governance, as well as to the problematic of the identity conflicts identified in the first part of this chapter.

More radically, in relation to redefining what is understood as "scale," some have questioned the two-dimensional character of what goes for space in political geography. In his call for a "vertical geopolitics," for example, Graham (2004) rightly observes a "flattening" of geopolitical discourse when it fails to consider the airspace associated with statehood (and its violation). Likewise, Alison Williams (2007, 2010) observes that the contemporary "battle space" of military activity calls for geographers to engage more with the aerial dimension of geostrategy and warfare. In fact, the dominance of land-based understandings of territory is even more problematic than simply in the aerial dimension, notwithstanding the fact that nation-states have long claimed sovereignty over other domains—from the "open seas" to the atmosphere (as in carbon trading, for example). The excessively narrow conception of territoriality held by political geographers has prevented them from considering the effective extension of sovereignty claims to outer space associated particularly with control over low earth orbit, which is at the very foundation both of contemporary U.S. "astropolitik" (MacDonald 2007; Sage 2008) as well as of telecommunications satellite networks that are one of the indispensable infrastructures of globalization (Verger 2002). Even the electromagnetic spectrum adds to the aspects of the overlooked "planetary commons" that certainly deserve more attention by political geographers in their attempt at reorienting the field away from the abstracted state to a fuller consideration of the broader conception of politics captured by the concept of scale.

One important preliminary before providing some examples of what is at stake in taking geographical scale more seriously is to define geographical scale. If cartographic scale refers to the scale of information or density of information on a map, geographical scale refers to the level of geographical resolution at which a given phenomenon is thought about, acted on, or studied. Conventionally, terms such as *local, national, regional*, and *global* are used to convey this meaning of scale. It is not the amount of information on a map, therefore, but the scale at which a particular phenomenon is framed geographically that matters. There is nothing very determinative about the terminology used to convey geographical scale of reference. The "local," for example, can be used to refer to areas of vastly different sizes and different character (urban, rural, etc.). Even the "global" may not mean worldwide but rather a geographical scope simply beyond the "continental." There is a certain arbitrariness to the terms, except that each only makes sense in relation to the others. So, within this constraint, geographical scale is imposed on the world and not inherent to it. Given that the terms are very much universal, however, it seems that today people pretty much everywhere tend to think about and organize themselves politically in reference to similar understandings of geographical scale.

The idea of geographical scale is analogous to the idea of "levels of analysis," in which one scale or another is regarded as key to explaining a given phenomenon. The doctrine of "reduction" presupposes that the lowest level is always the best, whereas "holism" presupposes that the whole is always greater than the sum of the parts, and hence that the most encompassing geographical scale is the best one. Reduction looks to isolated individuals, atoms, or neurological networks to explain human behavior. Holism looks to capitalism, culture, or the world-system. The idea of "emergence," however, suggests that many phenomena of interest in a field such as political geography cannot be adequately understood in terms of either reductionism or holism. Rather, phenomena are brought about across levels. In this view, only rarely is there a single scale at which total explanation can be found. From voting decisions to military strategy, the influences on what is done emanate across scales from the localized to the globalized, with a change in the "balance" of influences as nearby and distant influences fluctuate in their relative impacts.

In the past two decades there has been considerable debate on scale theory. The political-economic approach emphasizes scale as a social construct incorporating and expressing power in terms of the dominance of one scale over the other, from the global to the local. The poststructuralist approach emphasizes the relational character of scale and how scales come to be represented and enter into different discourses. The two perspectives are not necessarily mutually exclusive in their understandings of scale (and space). As Paasi (2004) has observed, relational networks and bounded territorial spaces may coexist. Therefore, the two approaches can actually complement each other (MacKinnon 2011), even if the former privileges direct and the latter intellectually mediated accounts of causation.

Four examples can serve to illustrate the importance of considerations of geographical scale in contemporary political geography. The first is the rising role of cities and their transnational networks in climate governance. The second is the role of territorially embedded networks of power in world politics. The particular case explored here is that of the al-Qaeda network in its association with the September 11, 2001, attacks in the United States. The third is the changing "balance" of geographical scales in electoral politics, explored here with reference to Italy between 1948 and 2008. The fourth example is of the emerging global system of finance linking major global cities and offshore banking centers in loosening the bonds of national financial regulation and currency markets and, consequently, undermining the close match between financial flows and national territories. In each case, however, it is the intersection of processes across scales rather the singular dominance of one scale (i.e., global networks versus states or the local) that is at work in the analysis.

Scale and Climate Governance

Climate change has traditionally been considered as a global problem requiring global solutions (Bulkeley 2010). Therefore, it has been hierarchically addressed through a centralized international regime that would bind every state to adopt carbon-reduction measures. While most of the attention so far has focused on international and national regimes, less consideration has been given to how such policies are implemented at the urban and regional scale. A dissatisfaction related to slow progress in the Kyoto Protocol's multilateral negotiations to limit greenhouse gas emissions has contributed to a blossoming of initiatives at the local scale in the 2000s, especially in the U.S., where the nonratification of the Kyoto Protocol left "a policy vacuum into which city and state authorities have ventured" (Bulkeley 2010, 239). According to the same author (2010, 236):

> Work on urban responses to climate change was among the first to challenge traditional approaches that regarded the international community and the development of regimes as the exclusive site of global environmental politics.

> Equally, scholars have recognized that the context within which urban actors are responding to the issue is critically shaped by the structures and processes of governing taking place at other scales and through multiple networks. Such multilevel approaches have been used to analyze the nature of climate change governance and its implications in terms of the reconfiguration of political authority within and beyond the state.

In fact, already in the 1970s and 1980s, cities in North America and Europe had started to engage with the prevention and control of environmental pollution, and since the 1990s a number of medium-size and small cities had pioneered policies focused on the issue of sustainable development and energy efficiency. In the past decade, a growing number of global cities and large metropolitan areas, also in the global South, have begun experimenting with a range of initiatives to address climate-change mitigation, within the context of transnational advocacy networks such as Cities for Climate Protection, the Climate Alliance, Energie-Cités, U.S. Mayors Climate Protection Agreement, the European Covenant of Mayors, the C40, and the Resilience Network.

Brenner (2004) has claimed that climate-change-oriented transnational networks can contribute to an overall rescaling of statehood. But Bulkeley (2010, 240) observes that it is not clear "the extent to which the state is being fundamentally reconfigured through this process or the governing of climate change is taking place through an alternative geography of authority and resources that operates around existing state practices." In fact, when transnational networks have tried to promote systematic policies on renewable energy, target setting, energy efficiency, or public transport, they have found that cities prefer to tackle these issues with greater autonomy. And in any case, with the exception of Japan and Sweden, the international climate governance agencies and national governments have done little to support such initiatives.

The pace of global urbanization—now including more than half of the world's population—indicates that cities do contribute a major quota of the world's carbon emissions. While the figures of the global contribution of cities to the increase in greenhouse gas emissions are disputed, the responsibility of each is obviously very different, according to factors like their size and economic development; for example, London's emissions are estimated to be equal to those of entire states such as Greece or Portugal. This implies that the role that such experiments could play both in climate-mitigation and -adaptation strategies is a major one, even if the assessment of the effectiveness of their measures is still quite limited. Moreover, as Dodman (2009) and Satterthwaite (2008) have pointed out, urban dwellers have lower per capita emissions than rural inhabitants, and within each city the emissions of single actors could greatly differ. Despite the many difficulties they face, however, urban climate policies and initiatives show how environmental challenges need to be addressed at a variety of scales. Cities' transnational networks are an increasingly diffused experiment that may open up to emerging forms of governance.

Scale and Terrorism

Terrorism—the targeting of civilians and military personnel for death and dismemberment by militant nonstate groups—is a political strategy originating in nineteenth-century anarchist attempts to unseat autocratic governments by assassination and social mayhem (Carr 2006). After World War II terrorism was adopted as a strategy by groups involved in colonial struggles for independence. Indeed, until the 1980s terrorism was almost always directed toward specific national governments and their territories by groups seeking to unseat those governments or break off part of the national territory for a new state. Classic examples would include the Stern Gang during Israel's war of independence; the IRA and ETA trying to force the British and Spanish governments, respectively, to unite Northern Ireland with the Irish Republic and break the Basque Provinces away from central rule; Palestinian terrorists engaged in a struggle against Israel; and various terrorist groups in Italy and Germany during the 1970s and 1980s. Some of these groups, of course, relied on financial support from sympathizers elsewhere. The IRA, for example, depended on funds collected among Irish Americans. The IRA and Palestinian groups received military support from various pariah states such as Gaddafi's Libya. But their goals were essentially nationalist, and the networks of support were usually national and, at most, international. Until September 11, 2001, many of the world's most deadly terrorist attacks still involved nationalist groups rather than transnational networks (table 5.1).

Many groups often characterized as "terrorist" are still largely national in orientation. The Palestinian Islamic movement Hamas, for example, is largely oriented toward bombing of Israeli civilian and military targets. The Zapatista movement in the Mexican state of Chiapas, frequently portrayed as a movement critical of globalization as represented by the NAFTA agreement between the U.S., Canada, and Mexico, is in fact a revolutionary group (and without any history of terrorist tactics) largely oriented to Mexican national politics and the demand for land reform in the Mexican south. Its use of Internet websites to attract worldwide attention is not enough to make it a global network. For that you also need both goals that extend beyond the boundaries of a single state and targets with global significance. The Zapatistas have shown evidence of neither.

But new groups that have gone in heavily for suicide bombings, taking their strategic cue from older Islamist groups with nationalist goals such as

Table 5.1. Terrorist attacks 2001–2010, by death toll and whether nationalist or not

Date	*Place*	*Death Toll*	*Nationalist or Not*
October 2001	Srinagar, Jammu, and Kashmir (India)	41	YES (Jaish-e-Mohammed)
December 2001	New Delhi, India	12	YES (Lashkar-e-Toiba)
October 2002	Bali, Indonesia	202	NO (Islamist)
October 2002	Moscow, Russia	160	YES (Chechen Islamist separatists)
December 2002	Grozny, Chechnya, Russia	83	YES (Chechen Islamist separatists)
May 2003	Russia	73	YES (Chechen Islamist separatists)
August 2003	Mumbai, India	71	YES (Lashkar-e-Toiba)
November 2003	Istambul, Turkey	57	NO (al-Qaeda)
March 2004	Madrid, Spain	191	NO (al-Qaeda)
May 2004	Grozny, Chechnya	10	YES (Chechen Islamist separatists)
August 2004	Moscow, Russia	11	YES (Chechen Islamist separatists)
September 2004	Beslan, Russia	334	YES (Chechen Islamist separatists)
September 2004	Jakarta, Indonesia	10	NO (Jemah Islamiyah)
December 2004	Jeddah, Saudi Arabia	9	NO (Saudi militants)
July 2005	London, UK	52	NO (al-Qaeda)
July 2005	Sharm el-Sheikh, Egypt	88	NO (Islamist)
August 2005	Bangladesh	2	NO (Islamist)
October 2005	Bali, Indonesia	20	NO (Islamist)
October 2005	Delhi, India	62	YES (Lashkar-e-Toiba)
November 2005	Amman, Jordan	60	NO (Islamist)
March 2006	Varanasi, India	28	YES (Lashkar-e-Toiba)
June 2006	Sri Lanka	68	YES (Tamil)
July 2006	Mumbai, India	209	YES (Lashkar-e-Toiba)
September 2006	Sri Lanka	92	YES (Tamil)
April 2007	Casablanca, Morocco	3	NO (Islamist)
May 2007	Hyderabad, India	16	NO (Islamist)
June 2007	Glasgow, London, UK	—	NO (Islamist)
December 2007	Algiers, Algeria	37	NO (Islamist)
May 2008	Jaipur, India	63	NO (Islamist)
October 2008	Guwahati, Assam, India	84+	YES (ULFA)
November 2008	Mumbai, India	173	YES (Lashkar-e-Toiba)
June 2009	Thailand, Philippines, Russia	27	YES (Islamist separatists)
July 2009	Grozny, Chechnya, Russia	6	YES (Chechen Islamist separatists)
July 2009	Majorca, Spain	2	YES (ETA)
August 2009	Nazran, Ingushetia, Russia	25	YES (Chechen Islamist separatists)

(*continued*)

Table 5.1. (*continued*)

Date	*Place*	*Death Toll*	*Nationalist or Not*
August 2009	Chechnya, Russia	8	YES (Chechen Islamist separatists)
October 2009	Pishin, Sistan-Baluchistan, Iran	31	YES (Jundallah)
November 2009	Bologoye, Russia	26	YES (Chechen Islamist separatists)
December 2009	Tokat, Turkey	7	YES (PKK)
January 2010	Nag Hammadi, Egypt	11	NO (Islamist against Christian Copts)
February 2010	Pune, India	16	YES (Lashkar-e-Toiba)
March 2010	Buenaventura, Colombia	6	YES (FARC)
March 2010	Moscow, Russia	52	YES (Chechen Islamist separatists)
March 2010	Kizlyar, Russia	12	YES (Chechen Islamist separatists)
April 2010	Ingushetia, Russia	2	YES (Islamist separatists)
April 2010	Isabela, Philippine	10	YES (Islamist separatists)
April 2010	Rangoon, Burma	9	YES (Antigovernment)
May 2010	Sirmak Province, Turkey	8	YES (PKK)
May 2010	Stavropol, Russia	8	YES (Chechen Islamist separatists)
June 2010	Aden, Yemen	13	NO (al-Qaeda)
June 2010	Turkey	18	YES (PKK)
July 2010	Kampala, Uganda	74	NO (al-Shabab)
July 2010	Zahedan, Sistan, Baluchistan, Iran	28	YES (Jundallah)
August 2010	North Ossetia, Russia	3	YES (Islamist separatists)
September 2010	Hakkari Province, Turkey	12	YES (PKK)
October 2010	Abuja, Nigeria	12	YES (MEND)
October 2010	Sanandaj, Iran	5	YES (Kurd separatists)
October 2010	Grozny, Chechnya	20+	YES (Chechen Islamist separatists)

Note: This selection does not include the conflict between Israel and Palestine, the civil war in Somalia, the war in Afghanistan, the 1,778 suicide bombings counted in Iraq between 2003 and May 2011, or around 40,000 victims of terrorist attacks between 2003 and 2010 in Pakistan.

Hamas (in Palestine), Islamic Jihad (in Egypt and Palestine), and Hezbollah (in Lebanon), have gone transnational. Al-Qaeda, for example, until recently headed by the Saudi millionaire Osama bin Laden, is charged with a long list of terrorist attacks directed toward the United States but largely oriented toward undermining the governments of a number of states in the Muslim world, particularly Egypt and Saudi Arabia, and establishing a modern version of the caliphate that prevailed throughout the Islamic world in the years following the death of the Prophet Muhammed. Osama bin Laden became involved in the Afghani jihad during the Soviet occupation of Afghanistan in the 1980s. After the Soviet defeat, he returned to Saudi Arabia. But as the U.S. militarized the Persian Gulf in the late 1980s, he decided to wage a global jihad against the United States. In Sudan after 1991, he started to build his Islamic army, networking with groups from Saudi Arabia, Egypt, Jordan, Lebanon, Iraq, Oman, Algeria, Libya, Tunisia, Morocco, Somalia, and Eritrea. He also began cooperating with extremists in Chad, Mali, Niger, Nigeria, and Uganda, and in Burma, Thailand, Malaysia, and Indonesia; supported the Moro Islamic Liberation front in the Philippines and the Pakistani insurrectionists in Kashmir; and offered assistance to Tajikistan Islamists. His network was also busy recruiting in the U.S. from bases in Brooklyn, Atlanta, Boston, Chicago, Pittsburgh, and Tucson. After moving back to Afghanistan, he established ties with the extreme Islamist Taliban government that came to power in Kabul in 1996 (Rashid 2000, Mishra 2000). This is where his agents came for training. Recruited from across the Arab diaspora in Western Europe and among disaffected young men all over, and funded by what was left of his part of the bin Laden family's fortune, other Saudi money, and contributions from Muslim groups in the United States and elsewhere, al-Qaeda consists of cells of activists taking advantage of the geographical mobility and social openness of Western societies and the new technologies of globalization (the cell phone, the Internet, international money wiring, etc.) to plan and carry out terrorist attacks in the hope of stimulating an anti-Muslim backlash in the United States, reprisals in the Islamic world that encourage recruitment of activists, and the withdrawal of American forces from the Arab world (Saikal 2000). Nevertheless, the threat from al-Qaeda has probably been systematically exaggerated by U.S. and other governments as a political mobilization tactic (Eilstrup-Sangiovanni and Jones 2008).

Al-Qaeda is undoubtedly a new type of political actor engaged in a new type of terrorism. Bin Laden was never a Che Guevara, the hero of the Cuban revolution and believer in spreading peasant revolution. Che operated country-by-country and never advocated terrorist acts against civilians. Radical chic in the 1960s did set great store by Che's internationalism, and he certainly had wealthy European backers, but his was a bottom-up rather than a top-down political strategy. He was also, we should remember, a socialist egalitarian, not a proponent of religious hierarchy and essentialist gender and religious distinctions. Al-Qaeda is worldwide in operation, bringing together operatives of diverse national origins in a loose global network, motivated to challenge the hegemony of Western capitalism and replace it with a puritanical version of political Islam throughout the Arab world (and elsewhere). Its terrorism is geared toward highly symbolic targets with no distinction drawn between civilian and military casualties.

Such targets are chosen to encourage recruitment of activists by creating a backlash when reprisals follow the spectacular acts of aggression. To a degree it also represents "blowback" on the policy of the Reagan administration in the 1980s of supporting radical Islamists against Soviet interests in Afghanistan and elsewhere (C. Johnson 2000). But al-Qaeda is not some sort of deterritorialized entity without connections to particular places. Its utopia rests on an imaginative geography of a restored caliphate. The United States is its target: a geopolitical abstraction seen as an earthly Satan. A distorted religious inspiration is fundamental to its goals and to its language. These are a mirror image of the idea of the "clash of civilizations" proposed by the American political scientist Samuel Huntington in 1993 (O'Hagan 2000). In this case an Islamic world is seen as in a death struggle with an infidel civilization represented by the United States, captain of the materialist West. The corruption and failure of states in the Islamic (particularly the Arab) world are put down entirely to cultural pollution emanating from the West. Only by expelling the West can the pollution be swept away.

There is a geographical embeddedness to al-Qaeda, therefore (Halliday 2001). It is not simply a disembodied network loose in a post-territorial world. What is difficult to grasp is that it is a cross-scale phenomenon: it operates across national boundaries and has transnational goals and appeal. It takes advantage of globalization to attack it, as its operations would not be possible but for the ease of communication and movement at a world scale that has emerged over the past forty years. Its geographical imagination is of a pan-Islamic world in which a strict interpretation of *sharia*, or Islamic law, will prevail and will be enforced by religious police such as those of Saudi Arabia. Its short-term goal, however, is the overthrow of a number of Arab governments by stimulating civil war. It can be adequately understood, therefore, only if placed in a multiscalar context.

One of the effects of the terrorist attacks of September 11, 2001, has been to revive the discourse of territorial danger and vulnerability in the United States that the end of the Cold War had seemingly undermined. Guarding the borders—homeland security—has once more become the leitmotif of American politics, with a heavy remilitarization of the country, a reinscription of the federal government as a centralized seat of power, a questioning of the traditional openness to foreign immigration, and a revived attachment to national symbols such as the flag and the pledge of allegiance. Yet the United States economy has become incredibly intertwined with that of the world at large and particularly with its closest neighbors, Canada and Mexico. Huge quantities of freight must cross U.S. borders every day for important sectors of the U.S. economy to function at all (Flynn 2002). The U.S. automobile industry, for example, depends heavily on parts shipped backward and forward across the two land borders. At the same time, the move toward "lean," "just-in-time" production, making items with parts shipped in shortly before use, puts a premium on minimizing delays at border crossings. The economic imperative thus comes up against the demand to improve border security, at all points of entry, notwithstanding the fact that the perpetrators of the attacks of September 11, 2001, entered the U.S. legally at major airports. The increased sense of territorial vulnerability mandates that the entire external boundary of the U.S. be reinscribed in political consciousness. Initiatives to treat borderland regions as distinctive zones with respect to economic development and citizenship policies, because of cross-border labor flows and political cross-allegiances, come under intense criticism. If symbolic of the emergence of a new type of global political actor such as al-Qaeda, therefore, September 11, 2001, also represents the reinstatement of the national boundaries as a fundamental element in American politics. Consequently, analysis of a single scale is not enough to understand what is going on. Although Osama bin Laden was killed in Abbottabad, Pakistan, on May 1, 2011, after an almost decennial search, al-Qaeda is still operative and has been recently reported to have adherents hiding in the African Sahel region in countries such as Niger, Mali, and Mauritania. Of course, we should not forget that perhaps fewer than ten thousand people across the world have been directly involved over the years as proponents of global terrorism on behalf of Muslim causes, suggesting that there is a danger of hyping the threat beyond proportion (Kurzman 2011).

Scale and Electoral Politics

Another reason why geographical scale must be taken more seriously in political geography is that the theoretical lens it offers provides a useful perspective on more conventional empirical topics in political geography such as the geographies of political movements, the politics of economic development, and electoral geography (see, e.g., Agnew 1997a; MacLeod and Goodwin 1999; Silvern 1999). This has become increasingly apparent as established political affiliations that had "nationalized" politics in many countries have eroded. For example, in the United States, presidential elections seem to have become increasingly regionalized compared to the 1970s and 1980s, with Democratic candidates doing best in the Northeast and on the West Coast and Republicans dominating in the continental interior and the South (recall the red state/blue state story from chapter 1). Of course, American national elections have always had a powerful regional or sectional cast to them. But this aspect seemed to be diminishing until recently. Likewise, in Britain, Canada, Italy, and France, established geographical patterns of support for political parties have been significantly disrupted by voter realignments.

In the final analysis, these are shifts in the votes of individuals from one party to another or none, or the result of new individuals entering the stock of voters and others leaving, either through migration or death. But voting does not take place in a social-geographical vacuum. People are subject to all sorts of influences from the state of their economic surroundings, the social groups to which they belong, and the cultural match between their lives and the appeals of politicians. These influences tend to covary across places. Parties also vary considerably in how well they are organized from place to place and, hence, how well they can communicate their messages to potential supporters. How all of these influences come together affects the geographical pattern of votes, with a localized pattern resulting from the highest degree of fragmentation of the electorate, a regionalized pattern occurring when the electorate takes on a greater degree of homogeneity across a contiguous area, and a nationalized pattern reflecting a widely shared set of preferences across all electoral districts (Agnew 2002).

At the end of World War II the Italian Communist Party (PCI) was one of the strongest in Europe. Due to the geographical position of Italy at the edge of the Iron Curtain, this was seen as a major risk within the Cold War's bipolar order. A communist government in Italy could shift the country outside of the Western alliance and into the Soviet sphere. For this reason strong pressure was placed on the center Christian Democratic Party (DC) (tied to the Vatican and with a strong territorial base through the local networks of the Catholic Church in Italy) from U.S. governments to ban the PCI from any Italian national government. After almost two decades of the DC ruling the country more or less alone, an opening to noncommunist left parties began in 1963 with the formation of center-left coalition governments. This increased ideological polarization in the country during the 1970s. Nevertheless, the U.S.-imposed ban on the PCI continued until the collapse of the Soviet Union, and only in the mid-1990s did former members of the now defunct PCI publicly disavow communism, start a new party, and eventually enter a national government. The memory of the "Red Scare" emanating from the possibility that the PCI would come to national government has become so deeply entrenched in Italian political discourse that even two decades after the end of the Cold War, leaders of the center-right coalition up to 2011 continued to use it to discredit the center-left.

Taking this historical context into account, Italian electoral politics since World War II can be interpreted in terms of three political-geographical "regimes" in which the places out of which Italy is made have had different degrees of electoral similarity and difference that have dominated in different periods. The first regime, dominant from 1947 until 1963, involved a regional pattern of support for the major political parties based upon place similarities that clustered regionally. The second, in effect from 1963 until 1976, witnessed the expansion of the PCI out of its regional strongholds into a nationally competitive position with the DC. This had different causes in different places, but the net effect was to suggest a nationalization of the two major parties. The third, characteristic of the period since 1976, has seen increased support for minor parties, including regional parties such as the Northern League (*Lega Nord*), the geographical "retreat" and political disintegration of the PCI and DC, and a more localized pattern of political expression in general, reflecting the increased "patchiness" of Italian economic growth and social change and the crisis of the system of existing parties after 1992.

For the period 1947–1963, Galli and Prandi (1970) divided the country into six zones based on levels of support for the three major parties—the PCI, the DC, and the Partido Socialista Italiano (PSI)—and strength of the major political subcultures, the socialist and the Catholic (see figure 5.6):

Zone I, the industrial triangle, covered northwest Italy in the 1950s (Lombardia, Piemonte, Liguria). Socialists, Christian Democrats, and Communists were all competitive in this region.

Zone II, *la zona bianca*, covered northeast Italy (Veneto, Trentino, and Friuli and some nearby provinces in Lombardia, such as Bergamo and Brescia). The DC was strongly entrenched in Veneto, and opposition was divided among a number of parties.

Zone III, *la zona rossa*, covered central Italy: Emilia-Romagna, Toscana, Umbria, and the Marche, extending to the provinces of Mantova and Rovigo to the north and Viterbo to the south. Within this area there were a few exceptions: Piacenza, Lucca (an "isola bianca"), and Ascoli Piceno. In this region the PCI was most strongly established in the countryside but also increasingly in the cities.

Zone IV, the south, included Lazio, Campania, Abruzzo e Molise, Puglia, Basilicata, and Calabria. This zone was historically the poorest and most marked by clientelistic politics. In the 1950s, the DC and the right-wing parties dominated but were faced with growing challenges from the PCI and PSI.

Zones V and VI, Sicily and Sardinia, had more complex political alignments than the peninsular south. For example, the PCI was well established in the southern provinces of Sicily (especially the sulphur-mining areas), while Sardinia had a strong regionalist party.

There were strongly rooted cultural "hegemonies" (party-based consensus building) in only two of these zones, *la zona bianca* and *la zona rossa*. However, in electoral terms, support for specific political parties was remarkably clustered regionally in 1953: the PCI in the center, the PNM (monarchists) and MSI (neofascists) in the south and Sicily, the DC in the northeast and south. In the 1950s, Italian politics followed a *regionalizing* regime, reflecting a similarity at the regional scale of place-based social, economic, and political relationships.

Figure 5.6. The electoral regions of Italy, c. 1963.
Source: Redrawn from Galli and Prandi (1970)

The second period, 1963–1976, marks a break with the regional pattern characteristic of the 1950s. Two electoral shifts were especially clear: the expansion of support for the PCI outside *la zona rossa* (along with its consolidation inside), particularly in the industrial northwest and parts of the south, and the breakdown of *la zona bianca* as a number of small parties made inroads in the previously hegemonic support for the DC in parts of the northeast. The net effect of these changes was a seeming *nationalization* of the major parties, even though they still maintained traditional areas of strength. These political changes were the fruit of the major economic and social changes Italy underwent in the late 1950s and early 1960s. A major expansion occurred in manufacturing and industrial employment, especially in the northwest, as a postwar "economic miracle" drew the Italian economy away from its predominantly agrarian base. At the same time that the industrial centers of the northwest were experiencing such dramatic economic and social change as a result of the economic boom and massive immigration, the rest of the country was experiencing shock waves emanating from the northwest. The extreme south (Puglia, Basilicata, and Calabria) was a major zone of emigration to the northwest and, with the exception of Taranto, without much industry. Where industry was established, it created pockets of new social and economic relationships in the midst of a rapidly depopulating rural society. In all these places and among immigrants from the south in the north, the PCI expanded its support in the late 1960s and early 1970s.

The other major feature of the period 1963–1976 was the so-called breakdown of the Catholic subculture or dominant position in *la zona bianca* and subsequent loss of DC voters. The DC had relied heavily on affiliated organizations, many of a religious nature, to mobilize its support. With the late 1960s heavy out-migration from rural areas in the northeast, and the growing industrialization in Venice, Treviso, Trento, and Pordenone, the traditional social networks and communal institutions upon which DC hegemony was based began to collapse. The nationalizing political-geographical regime peaked in 1976 when the DC and the PCI together accounted for 73 percent of the national vote. Although this trend had distinctive causes relating to the geographically differentiated social and economic impacts of the economic miracle and their interplay with political and organizational traditions, it was widely interpreted as a permanent nationalization of political life. The DC and the PCI were now national political parties.

The 1979 election, however, indicated a much more complex geography of political strength and variation than had been characteristic previously. Since then all parties have been less regionalized or nationalized than in the past. The 1983 and 1987 elections suggested a trend toward a *localization* or increased differentiation of political expression that has continued through the great political and electoral changes of the early 1990s. One cause was the increasingly differentiated pattern of economic change after a previous era of concentration. While the economic boom of the early 1960s concentrated economic growth increasingly in the northwest, by the late 1960s there was considerable decentralization of industrial activity out of the northwest and into the northeast and center. This new pattern of differentiated economic growth led some commentators to write of the "three Italies"—a northwest with a concentration of older heavy industries and large factory-scale production facilities; a northeast-center of small, family-based, export-orientated, and component-producing firms; and a still largely underdeveloped south, reliant on government employment but with some of the small-scale development (for example, in the vicinity of Bari or Caserta) characteristic of the "third Italy" (northeast-center). This terminology, though useful as a general characterization of a new economic geography, masks both a much more uneven and differentiated pattern at a local scale and the linkages between localized development and the big firms of the northwest. High concentrations of employment in major growth industries have, in fact, been widely scattered.

Other causes have also contributed to the localizing political-geographical trend from the late 1970s to the present. One was the failure of the parties to successfully adapt to social and economic change. In Trento and Udine (in the northeast), for example, the DC had problems adapting to the new economy. In large parts of the south and the northwest, the PCI was unable to capitalize on earlier successes mainly because, in the south, it had neither control over the state resources that lubricate the politics of many parts of that region, nor was it able to build a permanent following. In the northwest, its major vanguard of unionized workers was much reduced in economic importance at the same time that the other parties had become better organized and the particular problems of the southern immigrants, whom the PCI had previously recruited as voters, had largely receded from the political agenda. The emergence of effective regional-level governments in Italy as whole since 1970 also reinforced the localization of interests and sense of place. Where parties have achieved some strength and legitimacy through control over regional governments, they have been able to build local coalitions for national politics based upon the pursuit of local interests. The PCI benefited from its control of or participation in the regional governments of Emilia-Romagna, Toscana, and Umbria, but it suffered elsewhere. The former successes of the DC and the PCI in, respectively, *la zona bianca* and *la zona rossa* rested to a degree on the social institutions with which they were affiliated (unions, cooperatives, clubs, parishes, etc.), as well as social isolation. However, the shifting orientations of these institutions and the rise of the consumer society opened up possibilities for smaller parties. There is some evidence that, after the late 1960s, ties between the DC and the PCI and their supportive organizations, especially the unions for the PCI, had weakened. The parties themselves were responsible for some of this. In order to expand nationally, they often had to abandon or at least limit the ideological appeal that served so well in areas of traditional strength. More generally, parties do not always travel well. Thus, in comparing northeast with central Italy, the question of compatibility between "party style" and "local style" arises. As Stern (1975, 223) has noted in a comparison of *la zona bianca* in the northeast and *la zona rossa* in the center in the 1950s and 1960s:

> the evolution of two very different forms of political hegemony, each with distinct characteristics that necessitate sharply contrasting forms of maintenance. The Christian Democratic variety that flourishes in northeastern Italy is fueled efficiently by stable social organization that deemphasizes the place of politics in community life. In comparison the communist variant thriving in central Italy accents the urgent attention that political matters should command among the local citizenry and thereby constantly reaffirms the relatively recent sense of legitimacy that underlies PCI control.

Of course, these hegemonies always had local roots, and in some localities their power has been quite visible and persistent, as studies of Bologna and Vicenza suggest. Although, as Tesini (1986) suggests for Bologna, things *could* have turned out quite differently if local DC figures of a remarkably left-wing cast had achieved national office.

Finally, in 1992 the system of parties in place since the end of World War II came to an end. As a result of the end of the Cold War and disputes over the meaning and appropriateness of the term *communist*, the PCI had already regrouped as two new parties, the larger *Partito Democratico della Sinistra* (PDS) and the smaller hard-line *Rifondazione Comunista* (RC). As a result of investigation of systematic corruption in their operations, the PSI collapsed and the DC disintegrated into three separate parties, the *Partito Popolare* to the center-left and two smaller factions to the right. The fascist MSI was reborn as a new "postfascist" conservative party, *Alleanza Nazionale* (AN), and in 1994 a new party organized by the media tycoon Silvio Berlusconi, *Forza Italia*, attempted to replace the DC on the center-right. The proliferation of smaller parties continued. But a new electoral system, in effect for the first time in 1994, forced parties to look for coalition partners before elections. With the federalist/separatist *Lega Nord* a potent electoral force in many parts of northern Italy, all parties save *Forza Italia* and the PDS (by then known as DS for *Democratici di Sinistra*) were now largely local or regional in strength. Even these two parties must coalesce with some of the others to achieve national-government office. As a result of the local strength of different parties, support for all parties in 2001 is more obviously localized than it was in 1976 or in the 1950s.

Subsequent elections in 2001, 2006, and 2008 have seen a number of changes: in the electoral law, with the introduction of the *premio di maggioranza* or award of extra seats for winning the highest percentage of the votes; in the names of parties, with the DS changed into the *Partito Democratico* (PD) and *Forza Italia* and AN merged into the *Partito della Libertà* (PDL). In 2001 and 2008 the elections were won twice by Berlusconi allied to the

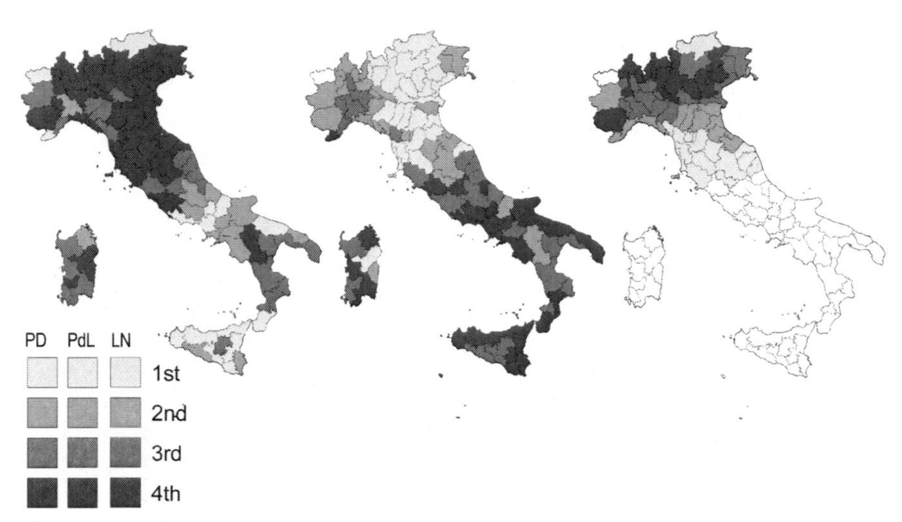

Figure 5.7. The preferences expressed in the 2008 Italian elections by decile for the three main parties (respectively PD, PDL, and LN) at the provincial scale.
Source: Authors.

Lega Nord, while in 2006 the PD alliance guided by Romano Prodi was able to defeat the coalition of the media tycoon. In general, recent elections have seen an increased regional polarization of the vote. In the last ones, most of northern Italy went to either the *Lega Nord* or the PDL, as the PD lost momentum but hung on to its dominance in central Italy (see, for 2008, figure 5.7). The South was crucial to the outcome, with the PD losing support in many places to the PDL. But there is also much localization/fragmentation within the regional pattern. The *Lega Nord* actually expanded its geographical range, at the expense of the PD in Piemonte, but by surpassing the PDL in the Veneto, with some signs of penetration in some provinces of Emilia-Romagna traditionally in the column of the PD, therefore gaining increased leverage over Berlusconi. The story of Italian electoral politics down the years, then, can only be told adequately if the intricate relationship between geographical scale and voting patterns is given central attention (Agnew 2002; Shin and Agnew 2008).

Scale and the New Global Finance

The "new economy" that helped bring about the localizing regime in Italian electoral politics after 1976 was itself a response to two major economic changes throughout the industrialized world in the 1970s with significant political origins: the shift in manufacturing production from large factories to small-scale production across a wider range of locations and the collapse of the Bretton Woods system of managed currency exchange rates in 1971

(Piore and Sabel 1984; Agnew and Corbridge 1995). If the former has underpinned to some degree the globalization of production across national boundaries, then the latter is partly responsible for the incredible expansion of a global financial system based on rapid interconnections between money markets and stock exchanges in major world cities (London, New York, Tokyo, Frankfurt, etc.) and the explosion of monetary flows between those cities and offshore banking centers devoted to easing the movement of capital from place to place. Government policies both nationally and as coordinated in international organizations such as the World Trade Organization (WTO), the World Bank, and the International Monetary Fund (IMF) have enabled these transformations. They would not otherwise have happened in the ways they have (Agnew 2012).

Absent the floating exchange rates between currencies that the collapse of the previous fixed-exchange-rate Bretton Woods system entailed, little of what became the rapidly growing and global financial system would have come about. Since 2007, of course, we have seen its downside: the near collapse of many banks in the U.S., Britain, Ireland, and Iceland because of their dabbling in incredibly risky financial products across multiple countries, particularly those involving variable interest rate mortgages provided to people who turned out not to have the means to pay them when house prices declined and interest rates went up (Preda 2009). As a concomitant to this, bailing out the banks created a public debt crisis as governments were faced with rapid increases in the debts they took on and increased difficulty in financing them. National governments proved insufficient to manage what had become a truly global financial system. In turn, however, this called into question the stability and utility of the global financial system.

At first sight, what would seem to augur a global economy based on financial flows would be a world in which place or territory and, hence, geographical scale "below" the global would matter progressively less and less. Surely this would be potent evidence for the emergence of a purely global phenomenon: the global financial network. A case in point is that of so-called e-cash, covering electronic debit and credit cards, various forms of smart cards, and true digital money. Many financial transactions still require mediation by banks and other institutions. But, increasingly, direct buyer-to-seller relationships cut out all intermediaries, as people communicate over the Internet and avoid banks, wholesalers, and retailers. In this world, new communication technologies seemed to have eliminated national borders and dissolved the link between income-producing activities and specific locations. Digitalization of finance was cutting money loose from its geographical moorings. Yet monetary regulation presumes that customers and institutions share a common space. The very idea of controlling the money supply, for example, assumes that national territory defines the scope of the market. In this context, therefore, many have spoken of the "increasing irrelevance of geographic jurisdiction in a digital world economy" (Kobrin 1997, 75). Not only does this increase the possibilities of fraud, money laundering, and financial scams, it also represents the diminished efficacy of governance rooted in mutually exclusive territorial jurisdictions. Indeed, the three powerful agencies most involved in rating stock and bond investment and thus the entire global financial system, the so-called credit-rating agencies, Moody's, Standard & Poor's, and Fitch, are private, profit-making firms that operate across the world's financial centers rather than in any manner according to territorial jurisdiction (see, e.g., Sinclair 2005).

This understanding of global finance, and globalization more broadly, has become widely accepted (Lévy 2008). But it is deeply problematic. In the first place, its effects have hardly been global or worldwide in coverage. They are by and large restricted to certain clearly demarcated regions of the world, largely consisting of certain centers in North America, Western Europe, and East Asia, with outliers elsewhere. The world after the Cold War has taken on a radically different geographical structure but not one in which any location is interchangeable with any other (see figure 5.8). In fact, the opposite has been the case. During the Cold War both sides had incentives to stimulate at least a veneer of economic development within their respective spheres of influence and in countries that they were courting for support in their struggle against their adversary (either the U.S. or the Soviet Union). Absent this political interference, market processes tend to reward regions and localities that have crucial advantages (in skilled labor, access to technology, and capital) over other ones. Richard Gordon (quoted in Castells 1996, 393) puts the matter succinctly:

> in this new global context, localized agglomeration, far from constituting an alternative to spatial dispersion, becomes the principal basis for participation in a global network of regional economies. . . . Regions and networks in fact constitute interdependent poles within the new spatial mosaic of global innovation.
>
> Globalization in this context involves not the leavening impact of universal processes but, on the contrary, the calculated synthesis of cultural diversity in the form of differentiated regional innovation logics and capabilities.

The net outcome is an increasingly uneven pattern of economic development at a world scale and increasing inequalities of income within countries as those places benefiting from the new economy (particularly high-tech industries, banking, insurance, and finance) prosper and others are left behind (e.g., Milanovic 2011). Politics in many countries has focused on improving relative global competitiveness but in so doing has favored

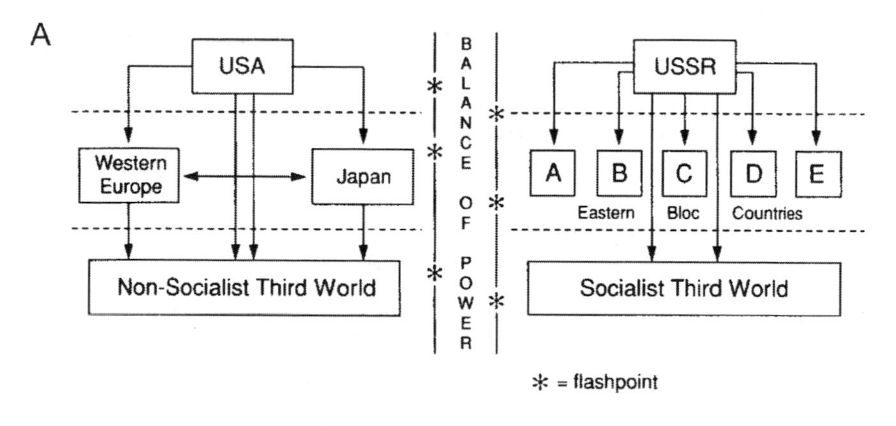

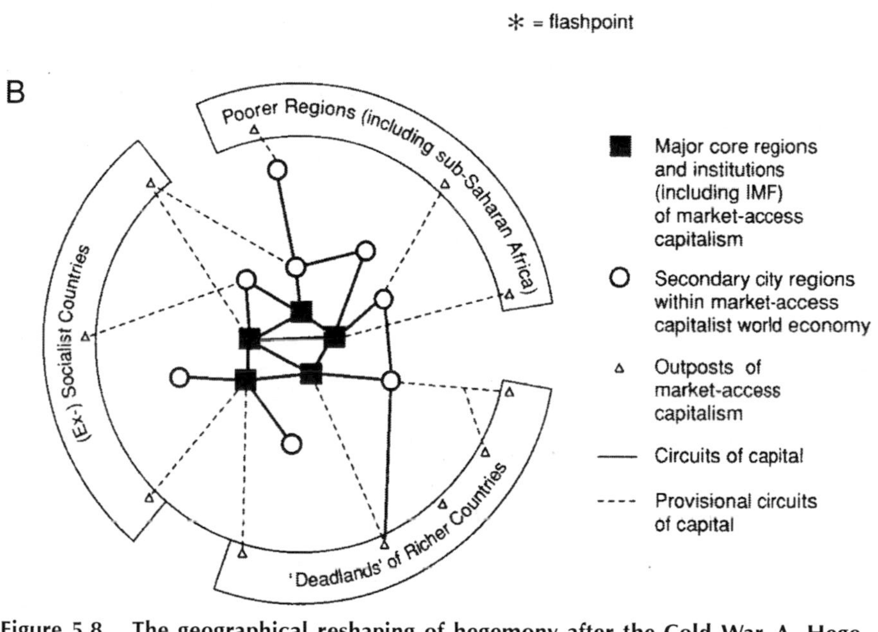

Figure 5.8. The geographical reshaping of hegemony after the Cold War. A. Hegemony of the Cold War geopolitical order. B. Hegemony of the post–Cold War geographical order.

Source: Agnew and Corbridge (1995, 206)

those groups best adapted to the "new world" and left others in the dust (e.g., Hacker and Pierson 2010).

In the second place, globalization is not without its own geopolitical roots. The globalization of financial markets is not the product of irresistible technological and market forces. In fact, states have been major sponsors of it for a whole variety of reasons: encouraging global financial centers as a strategy of economic growth (notably in the U.S. and British cases), satisfying financial and banking lobbies, facing difficulties in administering capital controls, and reacting to competitive pressures from other countries lowering barriers to capital mobility. The Bretton Woods Agreement had strongly endorsed the use of capital controls, but its abrogation by the United States in 1971 opened up the world economy to the possibility of an internationally competitive financial system, redolent of the period before the Great Depression of the 1930s. States reintroduced private global finance by giving freedom to market actors through liberalization of capital controls, coordinating with other states through collaboration between central banks, and choosing to limit controls on financial flows (Helleiner 1995).

This process was partly a response to the national economic difficulties facing some countries such as the U.S. and Britain in the 1970s and the perception that the state-interventionist approach no longer worked (Krippner 2011). But it was also the fruit of longer-term imperatives, particularly strong in the United States, to push not only new economic frontiers but also views of politics and cultural behavior beyond national shores into the world at large. In fact, much of the American ideology that was so important to the Cold War (U.S. democracy versus Soviet totalitarianism, U.S. liberal market versus Soviet planned economy, etc.) had a basis in U.S. history. The United States can in fact be seen as a prototype for the very processes underpinning globalization that began to achieve global proportions beginning in the 1970s: consumerism, financialization of the economy, privatization of pensions and other assets, and the "magic of the marketplace," among other factors (Agnew 2005).

But a third factor has also been important. States have hardly abandoned financial regulation to markets (Quinn 1997). Far from it. But the *variety* of ways in which states engage in regulation has allowed for increased "shopping around" between jurisdictions for the most auspicious fiscal and monetary conditions. It is the very division of the world into territorial jurisdictions, therefore, that has been a major stimulus to financial globalization. Indeed, "new places" have come into existence as centers of global finance precisely because they offer regulatory conditions that facilitate the movement of capital. Offshore financial centers have grown up to host banking, insurance, and other financial activities away from onshore regulatory authorities. Examples of such centers include the Bahamas and the Cayman Islands in the Caribbean, Gibraltar and Jersey in Europe, Bahrain in the Middle East, Singapore and Hong Kong in the Far East, and Vanuatu in the Pacific (Hudson 1998; Shaxson 2011). Beginning as tax havens for wealthy retirees or as poor areas with no alternative source of income, these places have become important nodes in global finance because they offer a haven from more rigorous onshore regulation but also because they are geared to the needs of the industries they serve without the political cross-pressures larger states must deal with. The different regulatory frameworks provided by different states, therefore, have had a powerfully formative influence on how global financial networks have developed.

Differences in material interests and conceptions of state action between states also set limits to the expansion of global networks. This is a fourth important caveat to the image of global finance running rampant over national boundaries. By way of example, the U.S. government and major financial interests have a particularly strong commitment to freeing financial markets that is not shared by many European and Asian states. One root of this lies in the greater reliance of European and Asian banks on interest income than U.S. banks, which receive far more of their profits from trading income, such as dealing in currency swaps and derivatives. European and Asian governments also have less faith in untrammeled markets, given the less positive economic growth profiles of Europe and Asia in the 1990s compared to the U.S. and the devastating impact of loosened capital controls on Asian economies in 1997–1998 (Wade 1998–1999). So even as the "globalist regime" in which the U.S., Britain, and some other countries have been most heavily invested has undergone wrenching shifts, other parts of the world, particularly China and Brazil, have shown greater resilience and look set to become relatively more important centers for the world economy as a whole in the years to come (Agnew 2009).

Finally, the image of an untrammeled global world formed by financial cyber-networks rests not so much on empirical evidence as on an enthusiasm for the world-without-boundaries prophesied since the beginning of human time. It is a kind of secular eschatology. It rests on the belief that connectedness with others and knowing them will automatically generate understanding and undermine conflict. Yet the globalizing world is an incredibly uneven and unequal one, as argued previously. What connectedness does is to make people in poorer places who have access to the new means of global communication increasingly aware of how different and materially poorer their lives are from those they come across on satellite TV or the Internet. At the same time, the absence of simple overarching conflicts (such as that of the Cold War) seems to have encouraged a turning away from the global or distant toward the local and national when it comes to news reporting in the U.S. and Europe (Moisy 1997; Archetti 2008). The opening up of global financial networks, therefore, is not paralleled by a surge in popular interest in the global. Ironically, for many people the local has become more important at exactly the time when the world as a whole has been compressed by the new communication technologies. Finding out about the impact of distant others, then, often encourages a turn homeward (see the next section on "Caring for Distant Strangers"). There are exceptions. Globalization has mobilized some people to fight against what they perceive as the depredations of global finance and production by organizing in transnational coalitions and disrupting the international meetings that symbolize its workings (G8 summits, IMF meetings, etc.) (see table 5.2). They are acting globally in defense of local and national interests and identities—as well as in defense of those they do not know elsewhere. Consequently, geographical scales must always be thought of relationally: no one scale ever exists in isolation from the others. Showing how scale intersects with finance both materially and discursively—and thus from a range of theoretical perspectives that will have to contend with one another in their ability to persuade others of their efficacy—will be a major element in the agenda of political geography in the years ahead.

Table 5.2. A world of protests: sites of major antiglobalization demonstrations, 1999–2010

City	*Country*	*Date(s)*
Seattle	USA	November 1999
Davos	Switzerland	January 2000, January 2001
Bangkok	Thailand	February 2000
Washington, DC	USA	April 2000, September 2002, October 2007, April 2009, October 2010
Salzburg	Austria	July 2000
Melbourne	Australia	September 2000, November 2006
Prague	Czech Republic	September 2000
Seoul	South Korea	October 2000
Porto Alegre	Brazil	January 2001, February 2002
Naples	Italy	March 2001
Montreal	Canada	April 2001
Honolulu	USA	May 2001
Barcelona	Spain	June 2001
Bologna	Italy	June 2001
Göteborg	Sweden	June 2001
Genoa	Italy	July 2001
New York City	USA	February 2002
Oslo	Norway	June 2002
Calgary and Ottawa	Canada	June 2002
Evian, Geneva, Lausanne	Switzerland	May–June 2003
Santiago	Chile	November 2004
Edinburgh	Scotland	July 2005
Hong Kong	China	December 2005
Hamburg	Germany	May 2007
Malmö	Sweden	September 2008
Athens	Greece	December 2008, May 2010
London	England	March–April 2009
Pittsburgh	USA	September 2009
Toronto	Canada	June 2010

NORMATIVE POLITICAL GEOGRAPHY

Political geography long avoided much concern about its presuppositions, such as determining the role of physical-environmental conditions in political organization, the historical-geographical origins of statehood, the "necessity" for empire, or the "obviousness" of territorial expansion, and their consequences for humanity. Two reasons for this lack of interest in the normative basis to political thought spring to mind. The first is the coincidence between the "rise" of political geography to both its classic and spatial-analytic manifestations and powerful pressures toward naturalized, empiricist-descriptive, and positivist-predictive modes of analysis. "Science," in a peculiarly narrow sense of the term, tended to rule understanding. Concern for the normative—the implicit judgments about what is "right" and "proper" in political arrangements—was viewed as subjective, ideological, and speculative in counterpoint to scientific analysis that hewed closely to the "facts" and saw theory as arising from observation rather than prior to it and hence conditioning of what constituted the "facts" at hand. From this point of view, the normative is purely a question of personal political preferences and without standing in what is presumed to be "objective" scholarship. This is how political geography was shut off intellectually from critical analysis of its concepts and presuppositions. The second reason for the absence of a normative political geography has been the close connection between the field and *raison d'état* or service to the state. Career prospects, such as memberships in prestigious national organizations and clubs and appointment at important universities, funding of research, and successful political influence, have been based on accepting the political status quo rather than opening it up for questioning. Thus, and unlike mainstream political theory's rigorous focus on concepts such as interest, power, and sovereignty, little or no attention has been paid to critical analysis of the concepts that the field has been based on: territory and territoriality, space and place, nation-states and statehood, nationalism, sovereignty and national identity, power and hegemony, geographical scale and networks, and violence and boundaries.

The theoretical aphasia finally began to dissipate in the 1970s with critiques of spatial analysis from broadly analytic-philosophical (Sack 1986) and Marxist (Harvey 1973) perspectives. But only in the 1990s did a critical attitude toward normative assumptions begin to permeate the field as a whole, extending from analysis of conventional views of state territoriality (Agnew 1994, 1999; Taylor 1994) to the language of foreign policy (Ó Tuathail and Agnew 1992), understandings of power (Allen 1999), the violence of national boundaries (Paasi 1995; Conversi 1999; Newman 2001), and place and political identity in democratic practice (Entrikin 1999).

Whatever else one might think about it, the so-called postmodern wave has had the effect of encouraging a much more critical outlook on the field and its practitioners, even though the postmodern attitude is itself cynical about the possibility of improving the human (or planetary) condition. In the 2000s, as geopolitical change offered a tsunami of reasons for critical inquiry, the debate has flourished, focusing on notions of empire (N. Smith 2003); hegemony (Agnew 2005); biopolitics and geopolitics (Legg 2011); sovereignty, globalization, and territory (Agnew 2009; Elden 2009); spaces of democracy (Barnett and Low 2004; Purcell 2008), and how "too much place" (in the sense of a lack of moral awareness brought on by limited social-geographical horizons) can allow injustice to flourish (Sack 2003).

Currently, four areas seem particularly important for the further development of a normative political geography, engaging with debates about actual changes in the world of politics and with important assumptions of political geography as a field as yet without the attention they deserve: transnational democracy, weapons and warfare, caring for distant strangers, and democratic states and intellectual freedom.

Transnational Democracy

The first area concerns arguments about "transnational democracy." In modern political thought, political space has been almost invariably associated with the idea of state territoriality; politics is about modes of government within and patterns of conflict and cooperation between the territories or tightly bounded spaces of modern states. Much thinking about democracy has been "trapped" by its orientation toward states and their presumed monopoly over political life. Plausibly, however, this rendition of the association between politics and place is both historically and geographically problematic. Not only is the state-territory relation a relatively recent one, it is one that has never completely vanquished other types of political geography (such as network-based kinship, and city-state or core-periphery imperial political systems) around the world (Agnew 1994). Writing about "failed" or "quasi" states in locations as diverse as East Africa, Central Asia, or Southern Europe, for example, often misses the fact that the absence of a working state bureaucracy throughout a given state's territory does not signify the absence of either politics or of alternative governance arrangements working locally or nonterritorially through networks of some sort or another, as happened, for example, with the "Islamic courts" of Mogadishu between the withdrawal of U.S. troops (1995) and the U.S.-supported Ethiopian invasion at the end of 2006.

From this point of view, "political space" cannot be reduced to state territoriality for two reasons. One is that states are always and everywhere challenged by forms of politics that do not conform to the boundaries of the state in question. For example, some localities have kinship or patronage politics, others have ethnic or irredentist politics oriented to either autonomy or secession, and others support political movements opposed to current constitutional arrangements, including the distribution of governmental powers between different tiers of government within the state. The second is that state boundaries are permeable, and increasingly so, to a wide range of flows of idea, investments, goods, and people that open up territories to influences beyond the geographical reach of current governmental powers.

To an increasingly vocal band of thinkers, state territory, society, and popular sovereignty are no longer congruent, and this undermines the quest for democracy at the national-state scale. To David Held (1999, 102), for example, one of the foremost advocates of cosmopolitan democracy, as the presumed autonomy of the territorial nation-state is undermined, so is the very basis to democracy:

> If democracy means "rule by the people," the determination of public decision making by equally free members of the political community, then the basis of its justification lies in the promotion and enhancement of autonomy, both for individuals as citizens and for the collectivity.

The only response must be to create new institutions beyond the state that match the increased geographic scope of the forces that are no longer effectively controlled by national governments. The form that such institutions might take can range from "scaled up" versions of American federalism (see Agnew 2001a) and enlarged "life worlds" (Habermas 1998) to the democratization of "global city regions" (A. J. Scott 1998).

Before accepting the logic of globalization as a new approach to democracy, however, a number of objections need to be investigated. The first one is that the deliberative nature of democracy requires a common ground or sense of place upon which to base a project of political equality and an allied commitment to common purpose. Territory has long served this vital purpose in democratic theory, providing a public space on which to anchor the abstract goals of democracy (Thaa 2001). Trying to nurture political participation that strengthens collective deliberation is difficult enough within state boundaries that have well-developed political institutions and some history of democratic practice; even a vibrant "civil society" does not necessarily guarantee democratic outcomes (Chambers and Kopstein 2001). Second, in democracy the public good necessarily remains contestable and subject to conflicting claims and arguments (Dahl 1999). For argumentation about the public good, however, a shared public sphere with effective communication is needed, which in turn requires a high degree of mutual understanding (e.g., Manin 1987). This is much more likely to occur within a contiguous population rather than across a diffuse network. Third, it is not clear that globalization has as yet undermined popular sovereignty (Yack 2001). The great landmarks in the history of popular sovereignty—from the English Glorious Revolution of 1688, the North and South American wars of independence, the French Revolution of 1789, the "springtime of the peoples" all over Europe in 1848 to the decay of the European colonial empires after World War II, the Soviet collapse of 1989–1992 and the "Arab Awakening" of 2011—are also major events in the history of democracy. This symbolism will not be easily displaced. Political identities are still significantly national ones, and democratic politics still rests significantly on national territories (e.g., Cerny 1999; C. Brown 2001). Fourth, the networks that advocates of transnational democracy point to as a possible basis for transcending the territorial parameters of conventional democratic theory have serious "legitimacy" problems. Nongovernmental organizations (NGOs), in particular, often claim to represent the interests of this or that group and in return channel funds, expertise, and information from donors to target groups (Hudson 2001). But to what degree can they legitimately make such a claim? To whom are the NGOs accountable? Whose views do they actually represent? What role should they play in the politics of the countries in which their operations are located? As yet, the ability of transnational networks to offer an alternative to conventional territorial forms of representation in even the most auspicious of circumstances is open to more than reasonable doubt.

Meanwhile, the political violence around the world that characterized the 2000s would seem to have at least temporarily eclipsed the hopes for transnational democracy and its attempt to reframe human activity beyond national and moral particularisms in terms of law, rights, and responsibilities, as far as human rights, principles of equal respect for the vital needs of human beings, or international environmental regimes are concerned. Still, as Held aptly recalls (2009, 536), the main cosmopolitan steps taken in the twentieth century that have altered state sovereignty in formal ways—from the UN system to the EU, from the laws of war to, more recently, the International Criminal Court—were "intitiated against the background of formidable threats to humankind—above all, Nazism, fascism and Stalinism . . . in the face of strong temptations to simply put up the shutters and defend the position of only some countries and nations."

The 2010s, therefore, augur to be a particularly appropriate time to rethink contemporary global governance exactly because of the difficulties the present international system has in coping with the increasing number of global challenges that "connect the fate of communities across the world" and need to be addressed on a transnational base, from environmental problems (climate change, biodiversity, water security, the safety of the food chain), to human security issues (poverty, conflict, and pandemic prevention), to regulatory issues (such as nuclear proliferation, toxic waste disposal, finance, tax and trade rules). As Held suggests, the possibility of enjoying multiple citizenships at different scales, in a similar way to how Europeans are now simultaneously citizens of the EU and of one of the twenty-seven member countries, could be a step in this direction, though even in Europe this is not without its tensions.

If the boundaries demarcating different scales of governance are often contested, as between the European and the national scales, as well as between administrative levels at the subnational scale, addressing the complexity of these boundaries remains preferable to geopolitical violence or market-based solutions that lead to dramatic increases in geographical inequality. Addressing the challenges to democracy at all scales, therefore, will be one of the main questions facing political geography in the years to come.

Weapons and War

Political power beyond state boundaries has often been seen as the projection of force by one state against another or others. Two ideas have long expressed this sentiment. One is the idea of anarchy in the space beyond the confines of one's ordered and domestic space that could only be managed by vigilant preparation for warfare. Without a substantial war machine prepared to strike at adversaries before they strike at you, a state is vulnerable to conquest and subjugation. The second one is the very commonly held view that states are in competition with one another for scarce resources and go to war with one another to wrest control of these from other states. That these ideas rest on longstanding features of human society, such as male fantasies of control and domination of others and the need to demonstrate individual prowess in warfare, are now widely recognized and thus subject to challenge and resistance (Enloe 1990; Goldstein 2001; Sharp 2002).

Beginning with the Cold War, however, the most militarily powerful states began showing reservations about using force against one another and, to a lesser extent, against weaker states. A reasonable inference to draw from this is that the orthodox assumptions no longer hold empirically as well as normatively (Deudney 1995; Kaldor 1999). This opens up the possibility of creating a world in which disputes between the most powerful states are less likely to lead to massive violence. The old moral question about the rightness of force as a solution to human conflicts takes on new meaning in a historical context in which the efficacy of interstate war is itself profoundly in question. Parenthetically, it is important to note that this trend does not necessarily portend a decline in total political violence. Indeed, there is a contemporaneous trend toward an increase in the prevalence of internal wars as state authority collapses in many states (from Colombia and Somalia to Afghanistan). This rather makes the point that state monopoly over the use of force is as increasingly problematic within state boundaries as it is beyond them.

From one point of view, therefore, the use of military force by powerful states against one another seems to be declining even if powerful states still intervene in weaker ones (Goldstein 2011). Three changes in military technology and two changes in the world economy seem crucial to this shift, notwithstanding inertial interests still committed to the production of weapons and the invention of threats to justify them in the United States and other major powers. The first military change is the impact of nuclear weapons. These have had the effect of not only introducing mutual deterrence but also of imprinting on potential combatants the likely escalation of all organized interstate violence into nuclear exchange. The unprecedented destructiveness of nuclear weapons and likely negative impact (through delayed radiation) on victors as well as the vanquished mean that their possessors paradoxically limit their military options by possessing them and discipline allies and adversaries alike by introducing the prospect of rapid escalation. Nuclear weapons also seem to favor defensive over aggrandizing military actions by raising the stakes for potential aggressors.

Even before the advent of nuclear weapons, however, a second feature of modern warfare had begun to erode the rational basis for its use. The economic and political costs of war between reasonably well-matched adversaries now exceed any conceivable collective benefit to national populations that can derive from it. There are of course domestic interests that are still served by war and preparation for it (weapon makers, military officers, etc.); only war now requires very costly investments that do not guarantee favorable results. The civil wars involving the intervention of the U.S. in Vietnam and the former Soviet Union in Afghanistan and the terrorist attacks on the United States in 2001 are reminders that even in apparently asymmetric conflicts the best armed need not prevail. Third, with respect to military factors, there is increasing revulsion among the world's most affluent populations with the human costs of modern large-scale war and the seemingly feeble benefits it generates. The use of military force faces a legitimation crisis. The loss of even a single pilot or soldier on the U.S. side now causes a total rethinking of American force commitments. This is perhaps a reflection of the increased visibility of the conduct of warfare in a visual age. Though televised war often takes spectacular or entertainment form, it also introduces an immediate sense of the deadliness of war that civilians in previous epochs of modern technological warfare never experienced (at least in the U.S.). At the same time there is disillusionment with the "fruits" of war. Gains often seem incommensurate with sacrifices.

The political inconclusiveness of many recent wars (such as that against Iraq in the Gulf in 1991 and the NATO "intervention" in Kosovo in 1999) adds to skepticism. The democratization of foreign-policy making in many countries has probably added to the questioning (Gelpi and Griesdorf 2001). Once reserved for small elites, foreign policy is now increasingly subject to public challenge and debate in ways unheard of forty years ago. The significance of public opinion and, hence, publicity in foreign- and domestic-policy arenas has increased worldwide. In this context, the pursuit of political change through nonviolent means has become increasingly possible and socially acceptable. In particular, successful nonviolent struggles (from Mahatma Gandhi's strategy of passive resistance in India in the 1920s and 1930s and Martin Luther King Jr.'s freedom marches for equal civil rights in the U.S. in the 1950s and 1960s to nonviolent opposition to Vietnam War, the "velvet revolutions" in Eastern Europe in 1989, and numerous movements for social and environmental change in contemporary India) and the 2011 regime changes in Tunisia and Egypt have made the case for sustainable political gains as a result of deliberately *not* resorting to violence (see, e.g., Bondurant 1958; Fairclough 1987; Tollefson 1993; Brinton and Rinzler 1990; Routledge 1994).

Finally, warfare has been far from a constant of modern human history. Recent studies suggest that if the twentieth century had the two most devastating wars, then earlier periods, particularly the eighteenth century, seem to have been relatively pacific. The lesson this historical record suggests is that wars between major powers are the ones to be avoided—at all costs (e.g., Hayes 2002). With little or no correlation with a whole series of putative "determinants" (arms spending, economic downturns, etc.), research on the global cycle of violence suggests that political agency is of prime importance in producing damaging wars, further implying that the devastating weapons that produce such outcomes can also be rejected. Witness the little-known Japanese decision to "give up the gun" in 1637 (Perrin 1979). Although particular features of early-modern Japanese society obviously played a role in abandoning a technology already in wide use after its introduction from Europe in 1543—the size and domination of the warrior samurai class whose chivalric code and swordmanship were threatened by the musket, lack of danger from foreign invaders with guns because of geographic isolation, attachment to swords as cultural artifacts, and general rejection of foreign ideas, such as Christianity and Western business practices, along with guns—the example does suggest that choices can be made that do not accept the inevitability of escalation in the adoption and use of military technology.

Paralleling these weapon- and military-related trends have been two trends related to the world economy. Interstate competition is now largely about capturing the benefits of global economic growth for one's territory more than about conquering another state's territory to capture its resources. Although the U.S. war with Iraq (2003) and the NATO intervention in Libya (2011) can be interpreted as such resource wars, their initiation and course suggests much complexity of motivation in both cases. Insertion into global corporate and financial networks now seems crucial to the course of national economic development. Exceptions help to show what is now largely the rule. Iraq's invasion of Kuwait in 1989–1990 was designed to capture that country's assets. What it revealed was the extent to which Kuwait's assets, other than its oil reserves, were mobile beyond the boundaries of the state. Indeed, the Kuwaiti government-in-exile contributed handsomely to the liberation of Kuwait from Iraqi occupation by a UN-sanctioned force through its continuing access to a large number of significant foreign investments from which it could pay for U.S. forces. At the same time, the main threats that increasingly face the world's most developed countries, such as the United States, are seen by significant commentators as flows of migrants that challenge national cultural homogeneity and hide potential terrorists rather than the traditional weapons threat from this or that state (except, perhaps, so-called rogue states, such as North Korea and Iran) (see Shapiro 1999; Ó Tuathail 2000a). The fear of migrants and refugees as a consequence of the 2011 war in Libya was initially a major issue in Europe, dividing Italy, France, Germany, and others over who should bear the burden and regulate the flow.

Second, technological changes have opened the possibility of escaping from the dilemma of competitive states chasing the same resources that might previously have led states into war with one another. In the present era of informational capitalism, perhaps with the exception of the energy industry, the most productive and profitable activities are no longer resource-intensive ones, such as heavy industries and extensive agriculture, but technologically intensive manufacturing, such as electronics and biotechnology, and service industries, such as tourism, finance, and personal services. These are best achieved by either generating external economies in local clusters of firms (as in California's Silicon Valley) or by tapping into global networks of specialized labor and customized production (e.g., A. J. Scott 1998). This is no longer a world in which territorially bigger is automatically economically or politically better. Hence, it is a world in which military force to achieve such rational goals as increased resources or to cope with the anarchy threatened by other states no longer seems to make as much sense as it once might have. In a world of rapid economic circulation, the rational link between territorial states and military force has become frayed, if not yet cut. In this context the ethics of weapons and warfare can begin to take on a more central role than before.

Caring for Distant Strangers

The Scottish philosopher David Hume once observed that ordinarily human empathy for others tends to diminish with distance as people consider first their nuclear and extended families, then the local community, followed by city or rural district, nation, geographical region, and finally the world beyond. In clan and intensely localized societies this may still be largely the case. As noted previously, the end of the Cold War has produced an increased parochialism in news gathering in North America and Western Europe even as the world has opened up to increased communication. But thanks to television, films, and the Internet, the world can now often seem as close as or even closer than our immediate neighbors. These technologies bring images of far-off suffering into our homes. They raise in a fundamentally new context the question raised by the Jewish Bible and the Christian Gospels: Am I my brother's/sister's keeper? Only now the question has a corollary: If I am my brother's/sister's keeper, who is my brother/sister? Or, in other words, if charity no longer just starts at home, where does it stop (see, e.g., Miller and Hashmi 2001)?

Political geography has conventionally accepted Hume's commonsense view, with an added emphasis on the nation as the fundamental unit of "caring." What might be some of the implications of adopting a less distance-decay-based understanding of the ethics of caring? The first implication is to extend the notion of caring from the "domestic" realm out into the world. If some feminist writers have associated an ethic of care with the familial and household realms and duties assigned to women in many societies, then extending the possible geographical scope of caring extends the political scope of caring as well (Robinson 1999). In the context of a so-called shrinking world, in which places are no longer adequately thought of as isolated spatial zones, two ethical responses have predominated. One has been to resurrect Immanuel Kant's idea of a single "humanity" as a reaction to a world both seemingly more united but also more threatening (environmentally, socially, etc.) and a cosmopolitan ethics (focusing on common rights and impartial justice) appropriate to it. A foundation of this perspective is the idea that life chances are fundamentally a product of where you are born and, hence, a matter of fate and chance rather than worth or merit. As Stuart Corbridge (1998, 37) puts it:

> To the extent that these Other people have been Us (the affluent), and to the extent that their lives are inextricably linked to our own, there are good reasons for attending to their needs and rights as fellow human beings in a manner that will make calls on our "own" resources and entitlements.

A second ethical response has been to argue that this vision of a common humanity is largely illusory. Rather, the contemporary situation is one in which interconnection coexists with deep differences. The globalizing world economy is one that is highly unequal—indeed, more unequal in many respects than that it has replaced. This means that caring cannot be simply about encouraging the powerful to "care about" the individual rights and humane treatment of others they do not know on a face-to-face basis, although it does not exclude this. It must be, rather, that responding morally to others is a capacity that is learned. This involves recognition that moral response is not a rational act of will but an ability to focus attention on another and to recognize the other as real. Such recognition is neither natural nor presocial, but rather something that emerges out of connections and attachments. In the context of (global) north-south relations, then, strategies would require sustained and continued attention to the lives, relations, and communities of people in developing countries, rather than to their individual rights or to the scope and nature of our obligations to them (Robinson 1999, 46–47).

A second implication of putting Hume's geography of care into question is the need to understand that others may have different moral discourses and practices than the ones "we" have. In other words, our caring must be aware of what matters to others. This does not necessarily mean endorsing moral relativism—I'm okay, you're okay—about, say, widow burning, clitorectomies, or child labor, but it does entail accepting the reality of a world of diverse moral belief and practice. David Smith (1999, 122–23) gives a number of nice examples of specific moral discourses, pointing out how in some languages there are no words equivalent to the ethics of recognition and responsibility that dominates Western understandings of morality, and how there exist very different orientations to self and the wider society. For example, in Mongolia moral discourse relies heavily on the use of heroic historical figures, rather than abstract rules, as guides to behavior and is oriented toward what is good for me (the self) rather than toward the sympathy for others found in other moralities (Humphrey 1997). Among the Fulani people in northern Nigeria, Smith—quoting Jacobson-Widding (1997)—explains:

> The three sins of lying, farting and stealing . . . exemplify the lack of self-control. The association is with shame rather than guilt: they are sins of commission, rather than omission in the sense of failing to meet the obligations or duties. To show self-mastery is a way of recognizing the relative position of people who interact, how the Fulani define themselves in relation to others, rather than their respective individual identities. It expresses collective social personhood. (1999, 123)

Moral cosmopolitanism can become moral imperialism if not sensitive to the myriad local moralities that are still alive and well around the world.

Democratic States and Intellectual Freedom

As political geographers have abandoned the cause of this or that state for self-consciously critical perspectives, they have not shown much interest in the social and political conditions that make this possible. If in the past political geography was simply the servant of states, what has changed to make the field both less state oriented and less beholden to state sponsorship? This transformation in the intellectual orientation of the field has coincided with the rising power of the United States and its academia, yet political geographers tend in general to be critical of the activities of its governments within and beyond American shores, as we have been at points previously. They often seem to take the side, if only through their one-sided critique of U.S. actions, of various despots and authoritarian regimes with whom U.S. governments have gone to war.

The contemporary making of political geography, therefore, has two seemingly contradictory features. On the one hand, the new political geography relies on a high degree of toleration and relative freedom of inquiry in the United States, France, Britain, India, Israel, Finland, and several other countries, in which political geography has prospered, that cannot be found much elsewhere in the world. What these countries have in common is a commitment, however fragile and intermittent it has been historically (witness the shameful treatment of Owen Lattimore in the United States during the "Red Scare" of the early 1950s described in chapter 4), to freedom of thought. In China and Saudi Arabia, to choose two other countries at random, state control over intellectual life is almost complete. Political dissent that is typical of many widely published academics in the contemporary United States or Britain would lead to prison or death in many other countries. That this is not simply a West versus East phenomenon is apparent when the crimes of German Nazism, Italian fascism, Soviet communism and Maoist communism in China are brought to mind. Political oppression is and has been widely dispersed geographically. On the other hand, however, we political geographers tend to be hypercritical of U.S. governments and their allies as agents of empire or hegemony, while often silently or openly endorsing various forms of tyranny elsewhere in the world. This is definitely not to say that a critical attitude toward the actions of U.S. governments should be abandoned. It is to say that this attitude should be extended to other actors and places as well (for exemplary cases that do precisely this, see Slater 1999; M. J. Watts 1999). There is a tendency to presume that the only political agency in the world is that of the United States and its allies, thus diminishing a priori the political agency of others. For example, the Gulf War of 1990–1991 was not initiated by the U.S. coalition that ejected Iraqi forces from Kuwait but by the prior invasion of Kuwait by Saddam Hussein's Iraq government. Now, one can discuss the colonial history of borders between Iraq and Kuwait as somehow inviting the Iraqi invasion, but it is simply disingenuous to suppose that the war was an entirely American affair in its genesis. Likewise, the 2011 NATO intervention in Libya followed a domestic uprising against the dictatorial Gaddafi regime. This could be seen as an opportunistic effort by the U.S.

and its allies to gain access to Libyan oil on better terms from a successor regime, but again it is simpleminded to see the war as being driven by that. It diminishes the agency of those on the ground locally and also tends to aggregate a wide array of interests and identities beyond those of the United States government.

Silence about one or other party matters enormously in accounts such as these. After September 11, 2001, who can doubt that the *idea* of the "clash of civilizations" or a fundamental cultural opposition between East and West, not necessarily geographical territories so much as clusters of ideas and images about reason, cities, social mobility, and the role of women (Buruma and Margalit 2002), is a two-way street to war rather than simply one emanating from the West? Phony as it might be in terms of the actual character of different societies, it seems to provide a simpleminded template for the zealous on both "sides." Bad ideas are not an American or Western monopoly.

From one point of view, *intellectual freedom* is a simple product of wealth sufficient to employ a group of people full-time as academics and intellectuals and of popular indifference to what they do as long as it does not undermine the wealth and power of the state providing the setting in which the intellectual freedom operates. This is a rather classic left-wing view of the "critical intellectual" in contemporary U.S. society. But not all wealthy countries produce critical intellectuals or invest much in higher education. So it is not simply a combination of wealth and indifference that provides the basis for the growth of critical and dissenting thought. This is rather a product of toleration and cosmopolitanism in the institutional settings in which such thought flourishes. Toleration is important because it allows for alternative conceptions of understanding and thinking *and for their propagation*. There must be a public sphere in which intellectuals can engage with one another. There are undoubtedly questions of access to this sphere. For example, there is a need to share a common language. This is not so much a problem with the natural sciences or mathematics, but it is a major barrier to inclusion in the humanities and social sciences. Command of English has certainly become a major constraint on access to an increasingly global intellectual sphere. Be that as it may, toleration provides the crucial social requirement for arguing your case against alternative ones. Absent this possibility, conventional understanding is never put up to the test. Cosmopolitanism is important because it presupposes a world larger than the one people typically inhabit on a daily basis and an associated sense of the need to positively accept and, indeed, value in some way or another the culturally heterogeneous form that this world takes. Even while being "products" of their cultural settings, critical intellectuals must be able to think beyond the "common sense" of the worlds in which they were raised. Absent this, they cannot legitimately claim the appellation "critical."

But neither toleration nor cosmopolitanism can be simply embraced independent of social-political context. There is a historical geography to both. Over the centuries, toleration and cosmopolitanism as prerequisites for intellectual freedom have "moved around." For example, while at one time the Arab world was characterized by both, this long ago ceased to be the case—although the recent opening signaled by the uprisings across the region offer some renewed grounds for hope. Today some sort of liberal democracy is a necessary, if insufficient, condition for both. Often, democracy as such gives an effective voice only to those in the majority. Resources, the valuing of liberal education, and safeguards against persecution are also necessary to its prosperity. The importance of liberal democracy to legitimizing and solidifying toleration is an uncomfortable fact for those of us who still find much to criticize in its practice. To accept toleration as a virtue, however, requires more than a set of material and institutional conditions conducive to it, including democracy. Richard Dees (1999) argues that groups must experience something akin to a "conversion" to accept toleration as a virtue. Comparing early-modern France and England, he sees this conversion as incontrovertibly contextual or geographical. For it to happen, he claims, groups with different beliefs must have the possibility to mingle with one another (not the least in the marketplaces of commerce), a common enemy helps to reduce the vitriol between them by creating an external foe (toleration rests to a degree on intolerance that is driven by fear), wide diversity between contending groups reduces the threat to each posed by any other, and increased individual autonomy loosens group ties and creates the possibility for alternative intellectual and political bonds. Without these contextually specific conditions, toleration will not take root.

Only in contexts where cultural identities are not under serious threat do tolerance and possibly cosmopolitanism seem to develop (see also the discussion of Gottmann in chapter 4). But cosmopolitanism and tolerance are not the same thing, even though some discussions seem to suggest as much (e.g., Walzer 1997). One can tolerate all manner of local groups and their ideas without accepting the main tenet of cosmopolitanism: the need to look outside one's own experience and identity in order to understand it. Exile groups, expatriates, migrants, and transients can be the most effective instigators of cosmopolitanism. They make a positive ethic out of cultural borrowing, but in order to borrow one must have a place. But if openness to other worlds and their ideas is one feature of the doctrine, another is the positive valuation of the settings in which the ideas are shared and reworked. The most important settings are large cosmopolitan cities (Ignatieff 1993). These have long been and remain the sites in which alternative worldviews engage and clash. It is out of this engagement and conflict that cosmopolitanism emerges. Around the world, however, cosmopolitanism seems a privilege reserved to those who live within the relatively secure boundaries of wealthy, developed states and who value the vibrancy of cultural diversity. Since the end of the Cold War these boundaries often seem less secure than they were. Elsewhere, particularly among those enamored of religious, neoliberal, or other fundamentalisms geared to this or that apocalypse or path to redemption, it excites fear and hatred from all those who share different iconographies. To them cosmopolitanism and toleration are equally despicable; symbols of hubris and deviation from the one true path. Such sentiments are alive and on the march around the world today, though they may take on different colors in different places. Ian Buruma and Avishai Margalit (2002, 4) provide an illustrative anecdote:

> There is a recurring theme in movies from poor countries in which a young person from a remote village goes to the big city, forced by circumstances to seek a new life in the wilder more affluent world. Things quickly go wrong. The young man or woman is lonely, adrift, and falls into poverty, crime, or prostitution. Usually the story ends in a gesture of terrible violence, a vengeful attempt to bring down the pillars of the arrogant, indifferent, alien city. There are echoes in this story of Hitler's life in Vienna, Pol Pot's in Paris, Mao's in Beijing, or indeed of many a Moslem youth in Cairo, Haifa, Manchester, or Hamburg.

Ironically, critical intellectuals (though rarely very cosmopolitan ones) have often been complicit in the rejection of the values of toleration and cosmopolitanism, finding much to praise in various forms of despotism and tyranny. Why should this be so? Having rejected *raison d'état*, they look for an inspiring alternative to the dull, mediocre, and psychologically oppressive regimes under which they live. This can be purely passive, refusing to identify sources of conflict in the world other than those they are immediately familiar with: those around them. But it can also be more active: what Mark Lilla has called "the lure of Syracuse" (Lilla 2001, 193–216). By this he means the temptation of the intellectual to follow the path of the philosopher Plato to the ancient Greek city of Syracuse (in Sicily) in the instruction of the tyrant Dionysus.

Although Lilla tends to give precedence to the "inner life" of intellectuals and their attraction to sponsorship by the powers-that-might-be, many of his case studies, from the Nazi celebrant Carl Schmitt and the Nazi sympathizer Martin Heidegger to the communist sympathizer Walter Benjamin and Alexandre Kojève and, finally, to the Nietzschean Michel Foucault and the "spiritual Marxist" Jacques Derrida, reveal three common features: an attraction to totalistic explanations putting faith in the explanatory power of a single force or factor—divine order, class struggle, authority or the *volk*, discourse, language *tout court*; an urge to impose this on everyone else, notwithstanding their own critiques of other intellectuals as dupes of the powerful; and the delusion that individual or collective identities are self-constructed in geographical isolation rather than constituted out of mutual contact and subsequent distanciation. The sad fact is that none of these avowedly *critical* intellectuals understood that their freedom to think and publicize their views depended on the very social-political conditions they criticized as abhorrent and in need of destruction. As Lilla (2001, xi) aptly makes the point:

> Fascist and Communist regimes were welcomed with open arms by many West European intellectuals throughout the twentieth century, as were countless "national liberation" movements that instantly became traditional tyrannies, bringing misery to unfortunate peoples across the globe. Throughout the century Western liberal democracy was portrayed in diabolical terms as the real home of tyranny—the tyranny of the capital, of imperialism, of bourgeois conformity, of "metaphysics," of "power," even of "language."

It is long past time for intellectuals to acknowledge and examine the social-political conditions that make their work possible and to guard against the "lure of Syracuse" by openly examining the roles of all of the actors they need to consider, not simply those nearby and well known whose crimes and misdemeanors may be more obvious but also perhaps less venal than those more distant and less familiar (Milosz 1955). Critically examining the political context of toleration and cosmopolitanism is a vital task for political geography in maintaining the field's own escape from the territorial trap of national parochialism in which it was long ensconced. If in the post–Cold War world the challenges are somewhat different from the past, not least because the world political map of territorial states is no longer the singular template for organizing politics that it once was, the intellectual terrain seems so much more varied and interesting.

CONCLUSION

In this chapter we have endeavored to identify and describe three relatively "new" areas in which political geography is currently being "made." From the politics of the environment through the articulation of ideas about the mediating role in politics of geographical scale to vigorous interest in the geographical aspects of normative political questions, political geography is currently under reconstruction. But this is not an entirely new phenomenon. As this book tries to show, political geography has been remade numerous times in order to match the changing times and geographies it is hoping to map. This is not to say that all is changing. Far from it: many of the older topics retain considerable interest, even as they are placed in new light. At one time, history dealt a bad hand to political geography. The rejection of its classic environmental determinism after World War II meant that the field went into long hibernation. During the Cold War the descriptive knowledge it could provide was not much in demand because the ideological nature of that conflict seemed to make places nothing more than chess pieces on a global game board. With the end of the Cold War the questions that political geography is concerned with have once more come center stage. Though the horizon beckons, the need is to have more people moving toward it, addressing the themes (and other, older ones) laid out here. The enterprise is open-ended as long as the world remains the complicated and dangerous place it is today.

One major difference with even the recent past is the complexity of the geopolitical context in which political geography is today being remade. This was the major theme of the first part of the chapter. Rather than a singular story or narrative such as can be established for previous eras, today no one theme supersedes all others. We draw attention to what seem to us to be some of the predominant features of global geopolitics from 1990 to the present. We identify these as an emerging interdependent world, identities in search of territories, the War on Terror, increased demands for democracy and self-government, and the crisis of the territorial model of political organization. Together these trends portend a world in which established theoretical perspectives and substantive empirical research alike are no longer as secure as they once seemed in the seemingly simpler times of the Cold War. The new themes we identify are ones that reflect the changed geopolitical context in which political geography will be made. Only the future will tell which or if any of these becomes the leitmotif of the times.

6

Conclusion

In a nutshell, the making of political geography has moved from attempting to explain politics, defined narrowly as the origins and workings of statehood and interstate conflicts, by natural causes or facts of nature to where today it attempts to understand the distribution and organization of power across geographical scales through the geographical imaginations, group and other affiliations, and the agency of people engaged in everyday struggles and conflicts. If this transformation in the meaning of political geography is the main theme of this book, it has been explained in terms of the conditioning effects of the changing geopolitical contexts that have provided the raw empirical material to which the field has had to continuously adjust its perspectives. By way of conclusion, we identify the four features of the field as a whole that have shown the most change and that have been raised in passing in previous chapters. Before doing that, we provide a brief statement about the general approach to the "making" of political geography taken by this book.

APPROACH

Although discussing many of the main authors and recognizably influential figures in the history of political geography as well as the field's central concepts, this book does not focus on the interior lives of authors or see the field as made entirely by intellectual influences, such as characterize the Great Men and history-of-ideas approaches to how fields of study develop and change. Rather, we have chosen to emphasize the importance of what we call the "geopolitical context" in the long-term evolution of the field. Partly because the subject matter of the field has always been strongly related to real-world events, in a much more direct way than say chemistry, physics, or, as some would have it, economics, it has had to find justification in its relevance to public and international affairs. At the same time, it has always been connected to the practice of politics, both actual and by aspiration, so it makes sense to think of making political geography in terms of the "times" in which research and writing have taken place.

We have also made a serious effort to show that where and by whom it has been undertaken has also had important consequences for the substance of what has been made. The movement in the intellectual center of gravity of the field, so to speak, from pre–World War I Germany to contemporary Britain and the United States has undoubtedly affected the shift away from the exclusive focus on the geography of statehood. The increased recruitment of people into the field from an increasingly wide range of national, social, sexual, gender, and ethnic backgrounds has also widened the empirical scope and theoretical range of the field as the discussion of ideas about "power" in chapter 2 made clear. Political geography was badly burned by environmental determinism. Although other determinisms have also threatened to overrun the field—the "economic" in the guise of political economy, for example—and environmental determinism has undergone something of a revival, the field has increasingly been defined in terms of a relatively autonomous "political." The critique of state-centrism since the 1990s has only favored this trend with the extension of concern into the manifestations of power operating in the nooks and crannies of everyday life.

But this external emphasis on the historical-geopolitical context has its limits. For one thing, there have always been individuals or schools of thought that go against the grain of the times, suggesting that authorship does count for something. The examples of Vidal and Reclus in the early years and Jean Gottmann somewhat later are cases in point. There is agency in intellectual life as there is in politics, even if it is subject to strong social influence. There is another caveat to note: not everything is about change. Waves of theory or differing political perspectives do not simply wash out previous ones. Once established, theoretical viewpoints and empirical research interests put down roots that continue to bear fruit for their partisans even as they are eclipsed in influence by newer ones. Struggling to accommodate to new conditions, they often appear anachronistic, as does environmental determinism in general or efforts at reading contemporary world politics from Mackinder's geopolitical model, but they remain at the very least as reminders of what were once widely accepted positions that could under certain social and geopolitical circumstances be brought back to life.

THE MAKING OF POLITICAL GEOGRAPHY

If the less exclusive attention to statehood is one of the most significant changes in focus for university-level political geography over the past one hundred years, what else has changed most importantly in the making of the field? We see four changes in the field as a whole as emerging out of the survey presented in previous chapters. This is not to suggest that there is a single disciplinary standard in ascendance. If anything, the opposite is the case. At the start of the field there was certainly a high degree of consensus around the determining role of the physical environment in political life. Later in the 1960s the doctrine of "scientific progress" suggested that over time a single core defined methodologically and theoretically (that of spatial analysis) would sweep over the field. But nothing has proved to be further from the truth. Rather than a monist practice, what we have achieved is an unruly pluralism of theories, methods, and subject matter. We have tried to show why this is what happened, particularly in the years since the 1970s with the onset of globalization and the ending of the Cold War.

The first change we would like to highlight is a slow and hesitant shift from the objectivist "view from nowhere," masking as it does definite social identities and political interests, to a more nuanced appreciation of the situatedness of knowledge creation and circulation. This is the idea that knowledge is always partial and biased, even as the aspiration remains that of convincing others of a "truthful" explanation. This recognition has been vital in allowing a more critical appreciation of the various phenomena studied in the field. The early years of political geography were less than intellectually stellar, at least by present standards, because the main practitioners hid their partisan objectives behind the veil of complete objectivity, using "facts of nature" to conceal the social, racial, and political agendas to which they were committed.

The second change has been a trend from naturalistic explanation in terms of environmental or social causes to an emphasis on political agency in historical-geographical contexts. In this respect, political geography evidences a return to pre-Enlightenment conceptions of the political as inherently related to the powers of human association, commitment, and institution building demonstrated in political discourse and action rather than the coercive effects of "structures" or "forces" pushing people to behave in this or that way. This behavioral understanding of power, common to statist, communitarian, and some liberal conceptions of the political (see chapter 2), is now challenged by views emphasizing human capability and performance more than human limits and deficiencies in the face of overriding forces, be they cultural, environmental, or economic.

The concepts of the field were once simply taken for granted and received next to no attention from its practitioners, who focused on developing empirical accounts of various "problems" associated with state borders, global strategy, or ethnic conflicts, for example. A third change has been the intense attention now given to the central concepts of the field—territory, boundary, governance, place, state, nation, nation-state, sovereignty, geopolitics, and so on. Not only do these concepts form the core of the field, such as it is, but most of them have also been neglected within the mainstream of political theory.

Finally, the field is increasingly characterized by an interest in historical contingency rather than causal explanation. By this we mean that historical periodization is seen as setting the limits to putative generalizations. Thus what might well have counted as a "successful" account of ethnic conflict during the Cold War years may not in the very different context of the years since the end of the Cold War. For example, during the Cold War the colonial past and the current prospects for exciting the interest of one or other of the superpowers were more likely to stimulate such conflict than they could possibly today. As times change, therefore, so should our explanations.

THE PARADOX OF POLITICAL GEOGRAPHY

There is still something of a paradox to the making of political geography from the 1890s to the twenty-first century. Even as the field as such has suffered from the vagaries of its reputation within the world of universities and the wider world, the subjects it studies have persisted or even revived in significance, even if in periods like that of the Cold War this was not always so obvious. Today, the subject matter is challenged by the idea of a new world in the making that "knows no boundaries." With the world on the way to becoming a giant pinhead in which where you are no longer counts for anything, connectivity, interdependence, global culture, and cyberspace are said to be displacing the bounded territorial spaces and grounded places that are the leitmotif of political geography. We hope that you see, from the everyday examples in chapter 1 of American politics, the so-called War on Terror, sectarian rioting in Belfast, the politics of disasters and famines; the more developed vignettes in chapter 2 about the global drug business, the territorial intractability of the Israel-Palestine conflict, ethnic separatism in Central Europe and the Caucasus, Somali piracy and the "failed state" of Somalia; and the other examples addressed in the body of the book that the pinhead world is a fantasy. The world is still packed with political dramas that can only be adequately understood if placed in their geographical contexts. This will remain the case until the world finally becomes a smooth spheroid with no resistance to human movement and without all of its current territorial divisions and inequalities. That's some way off.

The biggest challenge remains that beautifully related by Lorenzetti in his great fourteenth-century Sienese frescos discussed in the preface: how to understand (and manage) the conflict between good and bad government in city and countryside. The geographical means by which power is distributed and concentrated and the effects upon different people and places are as important today, if in different ways and involving different scales of operation, as they were in Lorenzetti's time. This makes the field as important as it ever was. That others rarely make the connection between the question of governance and the geography of power draws attention, however, to the persisting difficulty of getting others to recognize the field's intellectual significance.

Political geography has long been and remains an "edge" or marginal field in the university, in the sense that, at least since the days of Mackinder and Bowman, it has been distanced from the ears of the prince (the politically influential) and in the popular realm. Though it has also always had its subversives, such as Kropotkin and Reclus, as a field it has only recently in its own right become subversive or questioning of the intentions and designs of the powerful. Obviously, this can be seen as a weakness: who quotes political geographers on the national news? We should note, though, David Newman has been a longstanding columnist for the *Jerusalem Post*, suggesting how important a political-geographic perspective is in that setting as well as reflecting his capacity to challenge, provoke, and educate his audience. But edges can have their virtues. We need not invest in governmental regimes that then fail us. We can keep a healthy distance from the seats of power where we are open to new ideas rather than constantly recycling old ones that serve interests and identities other than the intellectual. In a poignant memoir, Tony Judt (2010, 206) captures in a very geographical way what is most at stake in being at the edge, and he does so by invoking the cosmopolitanism that we discussed in chapter 5 when he writes:

> I prefer the edge: the place where countries, communities, allegiances, affinities, and roots bump uncomfortably against one another—where cosmopolitanism is not so much an identity as the normal condition of life. Such places once abounded. Well into the twentieth century there were many cities comprising multiple communities and languages—often mutually antagonistic, occasionally clashing, but somehow coexisting. Sarajevo was one, Alexandria another. Tangiers, Salonica, Odessa, Beirut, and Istanbul all qualified—as did smaller towns like Chernovitz and Uzhhorod. By the standards of American conformism, New York resembles aspects of these lost cosmopolitan cities: that is why I live here.

Yet as Judt (2010, 207–8) subsequently notes, and writing of today's world literally and not metaphorically, we are still bedeviled by the divisions and animosities that always seem to undermine the cosmopolitanism (however conditional) that we favor:

> We are entering, I suspect, upon a time of troubles. It is not just the terrorists, the bankers, and the climate that are going to wreak havoc with our sense of security and stability. Globalization itself—the "flat" earth of so many irenic fantasies—will be a source of fear and uncertainty to billions of people who will turn to their leaders for protection. "Identities" will grow mean and tight, as the indigent and the uprooted beat upon the ever-rising walls of gated communities from Delhi to Dallas.
>
> Being "Danish" or "Italian," "American" or "European" won't just be an identity; it will be a rebuff and a reproof to those whom it excludes. The state, far from disappearing, may be about to come into its own: the privileges of citizenship, the protection of card-holding residency rights, will be wielded as political trumps. Intolerant demagogues in established democracies will demand "tests"—of knowledge, of language, of attitude—to determine whether desperate newcomers are deserving of British or Dutch or French "identity." They are already doing so. In this brave new century we shall miss the tolerant, the marginals: the edge people, my people.

Our cosmopolitan attempts at understanding this world in the making are more likely to help us manage them than becoming partisans of the very divisions involved would ever do. This is possibly the most important lesson from knowing something about the making of political geography. No single narrative can account for all the variety and diversity of the world. This is the virtue of theoretical pluralism. But understanding the ontological significance of differences for political geography can perhaps contribute to the difficult task of translating between them. It is also vital to the future making of political geography.

References

Agnew, J. A. (1987) *Place and Politics*. London: Allen & Unwin.

———. (1989) Beyond reason: Spatial and temporal sources of intractability in ethnic conflicts. In *Intractable Conflicts and Their Transformation*, ed. L. Kriesberg et al. Syracuse, NY: Syracuse University Press.

———. (1994) The territorial trap: The geographical assumptions of international relations theory. *Review* of *International Political Economy* 1:53–80.

———. (1997a) The dramaturgy of horizons: Geographical scale in the "reconstruction of Italy" by the new Italian political parties, 1992–95. *Political Geography* 16:99–121.

———. (1997b) *Political Geography: A Reader*. London: Arnold.

———. (1998) *Geopolitics: Re-Visioning World Politics*. London: Routledge.

———. (1999) The new geopolitics of power. In *Human Geography Today*, ed. D. Massey et al. Cambridge: Polity Press.

———. (2001a) The limits of federalism in transnational democracy: Beyond the hegemony of the US model. In *Transnational Democracy*, ed. J. Anderson. London: Routledge.

———. (2001b) The "view from nowhere" and the modern geopolitical imagination. In *On the Centenary of Ratzel's* Politische Geographie: *Europe between Political Geography and Geopolitics*, ed. M. Antonsich et al. Rome: Memorie della Società Geografica Italiana.

———. (2002) *Place and Politics in Modern Italy*. Chicago: University of Chicago Press.

———. (2003a) American hegemony into American empire? Lessons from the invasion of Iraq. *Antipode* 35:871–85.

———. (2003b) From Megalopolis to global city-regions? The political-geographical context of urban development. *Ekistics* 70 (418/419):19–22.

———. (2005) *Hegemony: The New Shape of Global Power*. Philadelphia: Temple University Press.

———. (2009) *Globalization and Sovereignty*. Lanham, MD: Rowman & Littlefield.

———. (2010) Waterpower: Politics and the geography of water provision. *Annals of the Association of American Geographers* 101 (3):463–76.

———. (2012) Putting politics into economic geography. In *Wiley-Blackwell Companion to Economic Geography*, ed. T. Barnes et al. Oxford: Wiley-Blackwell.

Agnew, J. A., and S. Corbridge. (1995) *Mastering Space: Hegemony, Territory, and International Political Economy*. London: Routledge.

Agnew, J. A., T. W. Gillespie, J. Gonzalez, and B. Min. (2008) Baghdad nights: Evaluating the US military "surge" using nighttime light signatures. *Environment and Planning A* 40:2285–95.

Agnew, J. A., and D. N. Livingstone, eds. (2011) *Sage Handbook of Geographical Knowledge*. London: Sage.

Albert, M. (2000) From defending borders towards managing risks? Security in a globalized world. *Geopolitics* 5:57–80.

Alcoff, L. (1991–1992) The problem of speaking for others. *Cultural Critique* 20:5–32.

Alexander, L. M. (1966) *World Political Patterns*. Chicago: Rand McNally.

Allen, J. (1999) Spatial assemblages of power: From domination to empowerment. In *Human Geography Today*, ed. D. Massey et al. Cambridge: Polity Press.

———. (2002) *Lost Geographies of Power*. Oxford: Blackwell.

———. (2003) Power. In *A Companion to Political Geography*, ed. J. Agnew, K. Mitchell, and G. Toal. Oxford: Blackwell.

Al Qassemi, S. (2011) He had it coming. *Arabian Business*, Thursday, May 5. http://www.arabianbusiness.com/he-had-it-coming-397887.html (accessed September 4, 2011).

Ancel, J. (1936) *Géopolitique*. Paris: Delagrave.

Anderson, P. (1974) *Lineages of the Absolutist State*. London: New Left Books.

Archer, K. (1993) Regions as social organisms: The Lamarckian characteristics of Vidal de la Blache's regional geography. *Annals of the Association of American Geographers* 83:498–514.

Archetti, C. (2008) News coverage of 9/11 and the demise of the media flows, globalization and localization hypotheses. *International Communication Gazette*. 70:463–85.

Badie, B. (1995) *La fin des territoires*. Paris: Fayard.

Bacevich, A. (2011) Tailors to the emperor. *New Left Review* 69 (May):101–24.

Bahadur, J. (2011) *The Pirates of Somalia: Inside Their Hidden World*. New York: Pantheon.

Bailey, I. (2008) Geographical work at the boundaries of climate policy: A commentary and complement to Mike Hulme. *Transactions of the Institute of British Geographers* 33 (3):420–23.

Balibar, E. (2002) World borders, political borders. *PMLA* 117 (1):68–71.

Barber, B. (1994) *Jihad versus McWorld*. New York: Ballantine.

Barnes, T. J., and J. S. Duncan, eds. (1992) *Writing Worlds: Text and Metaphor in the Representation of Landscape*. London: Routledge.

Barnett, C., and M. Low, eds. (2004) *Spaces of Democracy*. London: Sage.

Barnett, J. (2009) The prize of peace (is eternal vigilance): A cautionary editorial essay on climate geopolitics. *Climatic Change* 96 (1):1–6.

Bassett, K. (1999) Is there progress in human geography? The problem of progress in the light of recent work in the philosophy and sociology of science. *Progress in Human Geography* 23:27–47.

Bassin, M. (1987a) Imperialism and the nation-state in Friedrich Ratzel's political geography. *Progress in Human Geography* 11:473–95.

———. (1987b) Race contra space: The conflict between German *Geopolitik* and National Socialism. *Political Geography Quarterly* 6 (2):115–34.

———. (2003) Politics from nature. In *A Companion to Political Geography*, ed. J. Agnew, K. Mitchell, and G. Toal. Oxford: Blackwell.

Beaverstock, J. V., R. G. Smith, and P. J. Taylor. (2000) World-city network: A new metageography? *Annals of the Association of American Geographers* 90:123–34.

Beck, U. (1992) *Risk Society*. London: Sage.

Beer, G. (2000) *Darwin's Plots: Evolutionary Narratives in Darwin, George Eliot and Nineteenth-Century Fiction*. 2nd ed. Cambridge: Cambridge University Press.

Bennett, S., and C. Earle. (1983) Socialism in America: A geographical interpretation of its failure. *Political Geography Quarterly* 2:31–55.

Bentley, M. (1996) "Boundaries" in theoretical language about the British State. In *The Boundaries of the State in Modern Britain*, ed. S. J. D. Green and R. C. Whiting. Cambridge: Cambridge University Press.

Black, J. (2009) *Geopolitics*. London: Social Affairs Unit.

Blaikie, P. (1985) *The Political Economy of Soil Erosion in Developing Countries*. New York: Longman.

Blatter, J. K. (2001) Debordering the world of states: Towards a multi-level system in Europe and a multi-polity system in North America? Insights from border regions. *European Journal of International Relations* 7:175–209.

Blouet, B. W. (1987) *Halford Mackinder: A Biography*. College Station: Texas A & M University Press.

———. (2001) *Geopolitics and Globalization in the Twentieth Century*. London: Reaktion Books.

Boden, T., G. Marland, and R. J. Andres. (2010) *National CO2 Emissions from Fossil-Fuel Burning, Cement Manufacture, and Gas Faring: 1751–2007*. Oak Ridge, TN: Carbon Dioxide Information Analysis Center, Oak Ridge National Laboratory.

Boggs, S. W. (1940) *International Boundaries: A Study of Boundary Functions and Problems*. New York: Columbia University Press.

Bondurant, J. V. (1958) *Conquest of Violence: The Gandhian Philosophy of Conflict*. Princeton, NJ: Princeton University Press.

Bowman, I. (1921) *The New World: Problems in Political Geography*. Yonkers, NY: World Book.

———. (1942) Political geography vs geopolitics. *Geographical Review* 32:646–58.

———. (1945) The new geography. *Journal of Geography* 44 (6):213–16.

Brenner N. (2004) *New State Spaces: Urban Governance and the Rescaling of Statehood*. Oxford: Oxford University Press.

Breuilly, J. (1982) *Nationalism and the State*. Manchester: Manchester University Press.

Brinton, W. M., and A. Rinzler, eds. (1990) *Without Force or Lies: Voices from the Revolution of Central Europe in 1989–90*. San Francisco: Mercury House.

Brown, C. (2001) Borders and identity in international political theory. In *Identities, Borders, Orders: Rethinking International Relations Theory*, ed. M. Albert et al. Minneapolis: University of Minnesota Press.
Brown, L. (1981) *Building a Sustainable Society*. New York: Norton.
———. (1999) *State of the World*. New York: Norton.
Bruneau, M. (2000) De l'icone à l'iconographie, du religieux au politique, réflexions sur l'origine Byzantine d'un concept gottmannien. *Annales de Géographie* 616:563–79.
Bryner, G. C. (2000) *Environmental Movements in the Twentieth Century*. Lanham, MD: Rowman & Littlefield.
Bulkeley, H. (2010) Cities and the governing of climate change. *Annual Review of Environmental Resources* 35:229–53.
Burke, J. (2011) *The 9/11 Wars*. London: Allen Lane.
Buruma, I., and A. Margalit. (2002) Occidentalism. *New York Review of Books*, January 17, 4–7.
Butler, J. (1992) Contingent foundations: Feminism and the question of "postmodernism." In *Feminists Theorize the Political*, ed. J. Butler and J. W. Scott. New York: Routledge.
Calhoun, C. (1994) Social theory and the politics of identity. In *Social Theory and the Politics of Identity*, ed. C. Calhoun. Oxford: Blackwell.
Calleo, D. P. (1987) *Beyond American Hegemony: The Future of the Western Alliance*. New York: Basic Books.
Campbell, D. (1992) *Writing Security: United States Foreign Policy and the Politics* of *Identity*. Baltimore: Johns Hopkins University Press.
Caracciolo, L. (2011) *America vs America*. Roma-Bari: Laterza.
Carr, M. (2006) *The Infernal Machine: A History of Terrorism*. New York: New Press.
Carter, N. (2007) *The Politics of the Environment: Ideas, Activism, Policy*. New York: Cambridge University Press.
Castellano, L. (1991) *Il potere degli altri*. Florence: Hopeful Monster.
Castells, M. (1996) *The Rise of the Network Society*. Oxford: Blackwell.
Cerny, P. G. (1999) Globalization and the erosion of democracy. *European Journal of Political Research* 36:1–26.
Chambers, S., and J. Kopstein. (2001) Bad civil society. *Political Theory* 29:837–65.
Chartres, C., and S. Varma. (2010) *Out of water: From Abundance to Scarcity and How to Solve the World's Water Problems*. Upper Saddle River, NJ: Pearson.
Christison, K. (1999) *Perceptions* of *Palestine: Their Influence on US Middle East Policy*. Berkeley: University of California Press.
Claval, P. (1989) André Siegfried et les démocraties anglo-saxonnes. *Etudes Normandes* 38 (2):121–35.
———. (1994) From Michelet to Braudel: Personality, identity and organization in France. In *Geography and National Identity*, ed. D. Hooson. Oxford: Blackwell.
———. (1998) *Histoire de la géographie française de 1970 à nos jours*. Paris: Nathan.
———. (2000) *Hérodote* and the French Left. In *Geopolitical Traditions: A Century of Geopolitical Thought*, ed. K. Dodds and D. Atkinson. London: Routledge.
———. (2006) The scale of political geography: An historic introduction. *Tijdschrift voor Economische en Sociale Geografie* 97 (3):209–21.

Clout H. (2005), Elisée Reclus et nos géographies: An international conference at Lyon, 7–9 September 2005. *Journal of Historical Geography* 32 (2):441–43.
Clunan, A. L., and H. A. Trinkunas, eds. (2010) *Ungoverned Spaces: Alternatives to State Authority in an Era of Softened Sovereignty*. Stanford, CA: Stanford University Press.
Cohen, S. B. (1973) *Geography and Politics in a World Divided*. 2nd ed. New York: Oxford University Press.
Connolly W. (2002) *Neuropolitics: Thinking, Culture, Speed*. Minneapolis: University of Minnesota Press.
Conversi, D. (1995) Reassessing current theories of nationalism: Nationalism as boundary maintenance and creation. *Nationalism and Ethnic Politics* 1:73–85.
———. (1999) Nationalism, boundaries, and violence. *Millennium* 28:553–84.
Corbridge, S. (1994) Maximizing entropy? New geopolitical orders and the internationalization of business. In *Reordering the World: Geopolitical Perspectives on the Twenty-first Century*, ed. G. J. Demko and W. B. Wood. Boulder, CO: Westview.
———. (1998) Development ethics: Distance, difference, plausibility. *Ethics, Place and Environment* 1:35–53.
Cox, K. R. (1972) The neighborhood effect in neighborhood voting response surfaces. In *Models of Urban Infrastructure*, ed. D. C. Sweet. Boston: Heath.
———. (1973) *Conflict, Power and Politics in the City: A Geographic View*. New York: McGraw-Hill.
———. (1979) *Location and Public Problems*. Chicago: Maaroufa.
———, ed. (1997) *Spaces of Globalization: Re-asserting the Power of the Local*. New York: Guildford.
Cox, K. R., M. Low, and J. Robinson, eds. (2008) *The Sage Handbook of Political Geography*. London: Sage.
Cox, K. R., and D. R. Reynolds, eds. (1974) *Locational Approaches to Power and Conflict*. New York: Halsted.
Dahl, R. A. (1999) Can international organizations be democratic? A skeptic's view. In *Democracy's Edges*, ed. I. Shapiro and C. Hacker-Cordon. Cambridge: Cambridge University Press.
Dahlman, C., and G. Ó Tuathail. (2006) Bosnia's third space? Nationalist separatism and international supervision in Bosnia's Brčko district. *Geopolitics* 11:651–75.
Dalby, S. (2000) Geopolitics and ecology: Rethinking the contexts of environmental security. In *Environment and Security: Discourses and Practices*, ed. M. R. Lowi and B. R. Shaw. London: Macmillan.
———. (2002) Environmental geopolitics. In *Handbook of Cultural Geography*, ed. K. Anderson et al. London: Sage.
———. (2010). Recontextualising violence, power and nature: The next twenty years of critical geopolitics? *Political Geography* 29:280–88.
———. (2011) Welcome to the Anthropocene! Biopolitics, climate change and the end of the world as we know it. Paper presented at the annual meeting of the Association of American Geographers, Seattle, April 13 (session 2239).
Danielsson, S. K. (2009) Creating genocidal space: Geographers and the discourse of annihilation, 1880–1933. *Space and Polity* 13:55–68.
Davies, C. B. (1994) *Black Women, Writing and Identity: Migrations of the Subject*. London: Routledge.

Dees, R. H. (1999) Establishing toleration. *Political Theory* 27:667–93.

Demangeon, A., and L. Febvre. (1935) *Le Rhin: Problemes d'histoire et d'economie*. Paris: Armand Colin.

Derluguian, G. M., and S. L Greer, eds. (2000) *Questioning Geopolitics: Political Projects in a Changing World-System*. Westport, CT: Greenwood.

de Seversky, A. P. (1950) *Air Power: Key to Survival*. New York: Simon & Schuster.

Detienne, M. (2003) *Comment être autochtone*. Paris: Éditions du Seuil.

Deudney, D. (1995) Nuclear weapons and the waning of the real-state. *Daedalus* 124:209–51.

——. (2007) Anticipations of world nuclear government. Chap. 9 in *Bounding Power*. Princeton, NJ: Princeton University Press.

Diamond, J. (1997) *Guns, Germs, and Steel: The Fates of Human Societies*. New York: Norton.

Dijkink, G. (1996) *National Identity and Geopolitical Visions*. London: Routledge.

——. (2001) Ratzel's *Politische Geographie* and nineteenth-century German discourse. In *On the Centenary of Ratzel's* Politische Geographie: *Europe between Political Geography and Geopolitics*, ed. M. Antonsich et al. Rome: Memorie della Società Geografica Italiana.

Dillon, L. (2005) The State Department's Office of the Geographer: History and current activities. *Portolan* 64:35–45.

Dittmer, J., S. Moisio, A. Ingram, and K. Dodds. (2011) Have you heard the one about the disappearing ice? *Political Geography* 30:202–14.

Dodds, K. (2008) Icy geopolitics. *Environment and Planning D: Society & Space* 26:1–6.

——. (2010) Flag planting and finger pointing: The Law of the Sea, the Arctic and the political geographies of the outer continental shelf. *Political Geography* 29:63–73.

Dodman D. (2009) Blaming cities for climate change? An analysis of urban greenhouse gas emissions inventories. *Environment and Urbanization* 21:185–201.

Dottori, G. (2011) L'era dello Smart Power: il mondo post-occidentale di fronte alla rivoluzioni e al ritorno della politica di potenza. In *Nomos & Khaos: Rapporto Nomisma sulle prospettive economico-strategiche*, ed. Cucchi G. and G. Dottori. Bologna: Nomisma.

Dumont, L (1983) *Essais sur l'individualisme*. Paris: Seuil.

Duncan, J. S., and D. Ley, eds. (1993) *Place/Culture/Representation*. London: Routledge.

Duncan, N. (1996) Postmodernism in human geography. In *Concepts in Human Geography*, ed. C. Earle et al. Lanham, MD: Rowman & Littlefield.

Eakin, H. (2006) *Weathering Risk in Rural Mexico: Climatic, Institutional, and Economic Change*. Tucson: University of Arizona Press.

East, W. G., and A. E. Moodie, eds. (1956) *The Changing World: Studies in Political Geography*. Yonkers, NY: World Book.

Economist. (2011) Why the tail wags the dog. Emerging economies have greater heft on many measures than developed ones, August 6, 66.

Eilstrup-Sangiovanni, M., and C. Jones. (2008) Assessing the dangers of illicit networks: Why Al-Qaeda may be less threatening than many think. *International Security* 33 (2):7–44.

Elden, S. (2009) *Terror and Territory*. Minneapolis: University of Minnesota Press.

——. (2010) Land, terrain, territory. *Progress in Human Geography* 34 (6):799–817.

Elon, A. (2001) The deadlocked city. *New York Review of Books*, October 18, 6–12.

Engel, R. (2001) Inside Al-Qaeda: A window into the world of militant Islam and the Afghani alumni. *Jane's Defense News*, September 28.

Enloe, C. (1990) *Bananas, Beaches and Bases: Making Feminist Sense of International Politics*. Berkeley: University of California Press.

Entrikin, J. N. (1999) Political community, identity and cosmopolitan place. *International Sociology* 14:269–82.

Eva, F. (2001) For a Europe of flexible regions and not of region-states divided by ethnicity. *Geojournal* 52 (4):295–301.

Fairclough, A. (1987) *To Redeem the Soul of America: The Southern Christian Leadership Conference and Martin Luther King Jr*. Athens: University of Georgia Press.

Falah, G.-W. (2011) Epilogue: The challenge of keeping non-violent protest non-violent. *Arab World Geographer* 14 (2):179–87.

Farinelli, F. (2001) Friedrich Ratzel and the nature of (political) geography. In *On the Centenary of Ratzel's* Politische Geographie*: Europe between Political Geography and Geopolitics*, ed. M. Antonsich et al. Rome: Memorie della Società Geografica Italiana. Also with the same title in *Political Geography* 19 (2000):943–55.

Farish, M. (2010) *The Contours of America's Cold War*. Minneapolis: University of Minnesota Press.

Fearon, J. D., and D. D. Laitin. (1996) Explaining inter-ethnic cooperation. *American Political Science Review* 90:715–35.

Filiu, J.-P. (2011) *The Arab Revolution: Ten Lessons from the Democratic Uprising*. London: Hurst.

Finnegan, D. (2010) Darwin, dead and buried? *Environment and Planning A* 42: 259–61.

Florida, R. (2011) Globalization. In *Wiley-Blackwell Companion to Human Geography*, ed. J. Agnew and J. Duncan. Oxford: Wiley-Blackwell.

Flynn, S. E. (2002) America the vulnerable. *Foreign Affairs* 81 (1):60–74.

Fodor, J., and M. Piattelli-Palmarini. (2010) *What Darwin Got Wrong*. New York: Farrar, Straus and Giroux.

Forest, B. (1995) West Hollywood as symbol: The significance of place in the construction of a gay identity. *Environment and Planning D: Society and Space* 13:133–57.

Foucault, M. (1980) *Power/Knowledge*. Brighton: Harvester.

Foucher, M. (2011) *La bataille des cartes. Analyse critique des visions du monde*. Paris: Bourin.

Friedman, G. (2009) *The Next 100 Years*. New York: Doubleday.

Fukuyama, F. (1992) *The End of History and the Last Man*. New York: Free Press.

Gallagher, I. (1962) *Edmund Walsh S.J.: A Biography*. New York: Benziger.

——. (1982) *The Decline, Revival and Fall of the British Empire*. Cambridge: Cambridge University Press.

Galli, C. (2010) *Political Spaces and Global War*. Minneapolis: University of Minnesota Press.

Galli, G., and A. Prandi. (1970) *Patterns of Political Participation in Italy*. New Haven, CT: Yale University Press.

Gambi, L. (1994) Geography and imperialism in Italy: From the unity of the nation to the "new" Roman empire. In *Geography and Empire*, ed. A. Godlewska and N. Smith. Oxford: Blackwell.

Garmony, J. (2008) The spaces of social movements: o Movimento dos Trabalhores Rurais Sem Terra from a socio-spatial perspective. *Space and Polity* 12:311–28.

Garton Ash, T. (1999) Hail Ruthenia! *New York Review of Books*, April 22, 54–55.

Geiger, R. L. (1993) *Research and Relevant Knowledge: American Research Universities since World War II*. New York: Oxford University Press.

Gelpi, C. F., and M. Griesdorf. (2001) Winners or losers? Democracies in international crisis, 1918–94. *American Political Science Review* 95:633–47.

Geopolicity (2011) *The Economics of Piracy: Pirate Ransoms and Livelihoods off the Coast of Somalia*, May 2011, http://www.geopolicity.com/upload/content/pub_1305229189_regular.pdf (accessed September 8, 2011).

Gerhardt, H., P. E. Steinberg, J. Tasch, S. J. Fabiano, and R. Shields. (2010) Contested sovereignty in a changing Arctic. *Annals of the Association of American Geographers* 100 (4):992–1002.

Gilbert, E. (2011) Globalization. In *Wiley-Blackwell Companion to Human Geography*, ed. J. Agnew and J. Duncan. Oxford: Wiley-Blackwell.

Gilman, S. L. (1992) Plague in Germany, 1939/1989: Cultural images of race, space, and disease. In *Nationalisms and Sexualities*, ed. A. Parker et al. London: Routledge.

Gimpel, J. G., K. A. Karnes, J. McTague, S. Pearson-Merkowitz. (2008) Distance-decay in the political geography of friends-and-neighbors voting. *Political Geography* 27:231–52.

Goblet, Y.-M. (1934) *Le crépuscule des traités*. Paris: Berger Levrault.

Godlewska, A., and N. Smith, eds. (1994) *Geography and Empire*. Oxford: Blackwell.

Goguel, F. (1983) *Chroniques électorales . . . la cinquième république après de Gaulle*. Paris: Presses de la Fondation Nationale des Sciences Politiques.

Goldstein, J. (2001) *War and Gender*. Cambridge: Cambridge University Press.

———. (2011) *Winning the War on War: The Decline of Armed Conflict Worldwide*. New York: Dutton.

Gottmann, J. (1946a) French geography in wartime. *Geographical Review* 36:80–91.

———. (1946b) Soviet geography at war. *Geographical Review* 36:161–63.

———. (1947) De la méthode d'analyse en géographie humaine. *Annales de Géographie* 56:1–12.

———. (1952) *La politique des États et leur géographie*. Paris: Armand Colin.

———. (1961) *Megalopolis*. New York: Twentieth Century Fund.

———. (1966) Géographie politique. In *Géographie Générale-Encyclopédie de la Pléiade*. Paris: Gallimard.

———. (1973) *The Significance of Territory*. Charlottesville: University Press of Virginia.

———, ed. (1980) *Centre and Periphery: Spatial Variation in Politics*. London: Sage.

———. (1987) Siegfried, André. In *The Blackwell Encyclopedia of Political Institutions*, ed. V. Bogdanor. Oxford: Blackwell.

———. (1994) *Beyond Megalopolis*. Tokyo: Community Study Foundation.

Graham, H. D., and N. Diamond. (1997) *The Rise of the American Research Universities: Elites and Challengers in the Postwar Era*. Baltimore: Johns Hopkins University Press.

Graham, S. (2004) Vertical geopolitics: Baghdad and after. *Antipode* 36:12–23.
Gray, C. (1989) *The Geopolitics of Superpower*. Lexington: University Press of Kentucky.
Green, M. (2011) The killers of Karachi. *Financial Times*, August 16.
Greenfeld, L. (1992) *Nationalism: Five Roads to Modernity*. Cambridge, MA: Harvard University Press.
Gregory, D. (1989) Areal differentiation and post-modern human geography. In *Horizons in Human Geography*, ed. D. Gregory and R. Walford. London: Macmillan.
———. (2004) *The Colonial Present*. Oxford: Blackwell.
———. (2008). "The rush to the intimate": Counterinsurgency and the cultural turn in late modern war. *Radical Philosophy* 150:8–23.
———. (2011) The everywhere war. *Geographical Journal* 177:238–50.
Güney, A., and F. Gökcan. (2010) The "Greater Middle East" as a "modern" geopolitical imagination in American foreign policy. *Geopolitics* 15 (1):22–38.
Gurr, T. R. (2000) Ethnic warfare on the wane. *Foreign Affairs* 79 (3):52–64.
Habermas, J. (1998) Jenseits des Nationalstaats? Bemerkungenz u Folgeproblemen Derwirtschaftlichen Globalisierung. In *Politik der Globalisierungk*, ed. U. Beck. Frankfurt: Suhrkamp.
Hacker, J. S., and P. Pierson. (2010) *Winner-Take-All Politics*. New York: Simon and Schuster.
Halliday, F. (2001) *Two Hours That Shook the World: September 11, 2001; Causes and Consequences*. London: Saqi.
Halper, J. (2000) The 94 per cent solution: A matrix of control. *Middle East Report*, Fall.
Harris, C. D. (1997) Geographers in the US government in Washington DC during World War II. *Professional Geographer* 49:245–56.
Hart, D. M., and D. G. Victor. (1993) Scientific elites and the making of US policy for climate change research, 1957–74. *Social Studies of Science* 23:643–80.
Hartshorne, R. (1950) The functional approach in political geography. *Annals of the Association of American Geographers* 40:95–130.
Harvey, D. (1973) *Social Justice and the City*. Oxford: Blackwell.
———. (1982) *The Limits to Capital*. Chicago: University of Chicago Press.
———. (1983) Owen Lattimore: A memoir. *Antipode* 15:3–11.
———. (1989) *The Condition of Postmodernity*. Oxford: Blackwell.
———. (1993) Class relations, social justice and the politics of difference. In *Place and the Politics of Identity*, ed. M. Keith and S. Pile. London: Routledge.
———. (2010) *The Enigma of Capital and the Crises of Capitalism*. Oxford: Oxford University Press.
Hayes, B. (2002) Statistics of deadly quarrels. *American Scientist* 90 (1):10–15.
Heffernan, M. J. (1994) The science of empire: The French geographical movement and the forms of French imperialism, 1870–1920. In *Geography and Empire*, ed. A. Godlewska and N. Smith. Oxford: Blackwell.
———. (1998) *The Meaning of Europe: Geography and Geopolitics*. London: Arnold.
Heidegger, M. (1959) *An Introduction to Metaphysics*. New Haven, CT: Yale University Press.
Held, D. (1999) Democracy and globalization. In *Democracy's Edges*, ed. I. Shapiro and C. Hacker-Cordon. Cambridge: Cambridge University Press.

——, ed. (2000) *A Globalizing World? Culture, Economics, Politics*. London: Routledge.

——. (2009) Restructuring global governance: Cosmopolitanism and the global order. *Millennium—Journal of International Studies* 37 (3):535–47.

Helleiner, E. (1995) Explaining the globalization of financial markets: Bringing the state back in. *Review of International Political Economy* 2:315–41.

Henrikson, A. K. (1980) America's changing place in the world: From "periphery" to "centre"? In *Centre and Periphery: Spatial Variation in Politics*, ed. J. Gottmann. London: Sage.

Hepple, L. W. (1986) The revival of geopolitics. *Political Geography Quarterly* 5:S21–S36.

——. (2000) *Geopolitiques de Gauche*: Yves Lacoste, *Herodote* and French radical politics. In *Geopolitical Traditions: A Century of Geopolitical Thought*, ed. K. Dodds and D. Atkinson. London: Routledge.

Herbert, S. (2006) *Citizens, Cops, and Power: Recognizing the Limits of Community*. Chicago: University of Chicago Press.

——. (2011) Segregation. In *Wiley-Blackwell Companion to Human Geography*, ed. J. A. Agnew and J. S. Duncan. Oxford: Wiley-Blackwell.

Herwig, H. H. (1999) *Geopolitik*: Haushofer, Hitler and *Lebensraum*. *Journal of Strategic Studies* 22 (2–3):218–41.

Hobsbawm, E. J. (1990) *Nations and Nationalism since 1780: Programme, Myth and Reality*. Cambridge: Cambridge University Press.

——. (1994) *Age of Extremes: The Short Twentieth Century*. London: Michael Joseph.

Homer-Dixon, T., and J. Blitt, eds. (1998) *Ecoviolence: Links among Environment, Population, and Security*. Lanham, MD: Rowman & Littlefield.

Hooson, D. (1964) *A New Soviet Heartland*. New York: Van Nostrand.

——. (1966) *The Soviet Union: Peoples and Regions*. Belmont, CA: Wadsworth.

——. (1984) Continuity and change in Soviet geographical thought. In *Geographical Studies of the Soviet Union: Essays in Honor of Chauncy D. Harris*, ed. G. Demko and R. Fuchs. Geography Department Research Papers 211. Chicago: University of Chicago.

Howard, D. (2010) *The Primacy of the Political: A History of Political Thought from the Greeks to the French and American Revolutions*. New York: Columbia University Press.

Howitt, R. (2003) Scale. In *A Companion to Political Geography*, ed. J. Agnew, K. Mitchell, and G. Toal. Oxford: Blackwell.

Hubert J.-P. (1998) À la recherche d'une géométrie de l'espace habité chez C. Vallaux, J. Gottmann et G. Ritchot. *L'Espace Géographique* 3:217–27.

Hudson, A. C. (1998) Reshaping the regulatory landscape: Border skirmishes around the Bahamas and Cayman offshore financial centers. *Review of International Political Economy* 5:534–64.

——. (2001) NGOs' transnational advocacy networks: From legitimacy to political responsibility? *Global Networks* 1:331–52.

Hugill P. J. (1999) *Global Communications since 1844: Geopolitics and Technology*. Baltimore: Johns Hopkins University Press.

Hulme, M. (2008) Geographical work at the boundaries of climate change. *Transactions of the Institute of British Geographers* 33 (1):5–11.

——. (2009a) Governing and adapting to climate. A response to Ian Bailey's commentary on "geographical work at the boundaries of climate change." *Transactions of the Institute of British Geographers*, n.s., 33:424–27.

——. (2009b) *Why We Disagree about Climate Change: Understanding Controversy, Inaction and Opportunity*. Cambridge: Cambridge University Press.

——. (2010) Moving beyond climate change. *Environment* 52 (3):15–19.

——. (2011) Reducing the future to climate: A story of climate determinism and reductionism. *Osiris* 26 (1):245–66.

Humphrey, S. (1997) Exemplars and rules: Aspects of the discourse of moralities in Mongolia. In *The Ethnography of Moralities*, ed. S. Howell. London: Routledge.

Huntington, S. P. (1993) The clash of civilizations? *Foreign Affairs* 72 (3):22–49.

IAEA. (2010) *International Status and Prospects of Nuclear Power*. GOV/INF/2010/12-GC(54)/INF/5. www.iaea.org/About/Policy/GC/GC54/GC54InfDocuments/English/gc54inf-5_en.pdf (accessed September 1, 2011).

Ignatieff, M. (1993) *Blood and Belonging: Journeys into the New Nationalism*. New York: Noonday Press.

Iseri, E. (2009) US grand strategy and the Eurasian Heartland in the twenty-first century. *Geopolitics* 14:26–46.

Jacobson-Widding, A. (1997) "I lied, I farted, I stole . . .": Dignity and morality in African discourse on personhood. In *The Ethnography of Moralities*, ed. S. Howell. London: Routledge.

Johnson, C. (2000) *Blowback: The Costs and Consequences of American Empire*. New York: Henry Holt.

Johnson, N. (1995) Cast in stone: Monuments, geography, and nationalism. *Environment and Planning D: Society and Space* 13:51–65.

Johnston, R. J. (1990) Lipset and Rokkan revisited: Electoral cleavages, electoral geography and electoral strategy in Great Britain. In *Developments in Electoral Geography*, ed. R. J. Johnston et al. New York: Routledge.

——. (2003) Territory and territoriality in a globalizing world. *Ekistics* 70 (418/419):64–70.

Jones, R. (2011) Border security, 9/11 and the enclosure of civilization. *Geographical Journal* 177:213–17.

Jones, S. B. (1954) A unified field theory of political geography. *Annals of the Association of American Geographers* 44:111–23.

Jones, S. H., and D. B. Clarke. (2006) Waging terror: The geopolitics of the real. *Political Geography* 25:298–314.

Judkins, G., M. Smith, and E. Keys. (2008) Determinism within human–environment research and the rediscovery of environmental causation. *Geographical Journal* 174 (1):17–29.

Judt, T. (2010) *The Memory Chalet*. New York: Penguin.

Kaldor, M. (1999) *New and Old Wars: Organized Violence in the Global Era*. Cambridge: Polity Press.

Kaplan, R. D. (2010) *Monsoon: The Indian Ocean and the Future of American Power*. New York: Random House.

Karanian, M. (2000) The Karabagh story. *American Philatelist* 114 (3):264–68.

Kasperson, R. E., and J. V. Minghi, eds. (1969) *The Structure of Political Geography*. Chicago: Aldine.

Kaufmann, C. D. (1998) When all else fails: Ethnic population transfers in the twentieth century. *International Security* 23:120–56.

Kennedy, P. (1987) *The Rise and Fall of the Great Powers: Economic Change and Military Conflict from 1500 to 2000*. New York: Random House.

Kern, S. (1983) *The Culture of Time and Space, 1880–1918*. Cambridge, MA: Harvard University Press.
Kilcullen, D. (2009). *The Accidental Guerilla: Fighting Small Wars in the Midst of a Big One*. New York: Oxford University Press.
Kirby, A. (1994) What did you do in the war, Daddy? In *Geography and Empire*, ed. A. Godlewska and N. Smith. Oxford: Blackwell.
Klare, M. (2001) The new geography of conflict. *Foreign Affairs* 80 (3):49–72.
Klemencic, M. (1999) Kosovo: The Albanians in the "cradle of the Serbian people." *International Boundaries and Security Bulletin* 6 (4):52–61.
Kobrin, S. J. (1997) Electronic cash and the end of national markets. *Foreign Policy* 107:65–77.
Kofman, E. (2008) Feminist transformations in political geography. In *Sage Handbook of Political Geography*, ed. K. R. Cox et al. London: Sage.
Krauthammer, C. (1990) The unipolar moment. *Foreign Affairs* 70 (1):22–33.
Krippner, G. R. (2011) *Capitalizing on Crisis: The Political Origins of the Rise of Finance*. Cambridge, MA: Harvard University Press.
Krishna, S. (1993) The importance of being ironic: A postcolonial view of international relations theory. *Alternatives* 18:385–417.
———. (1994) Cartographic anxiety: Mapping the body politic in India. *Alternatives* 19:507–21.
Kühl, S. (1994) *The Nazi Connection: Eugenics, American Racism, and German National Socialism*. New York: Oxford University Press.
Kurti, L. (2001) *The Remote Borderland: Transylvania in the Hungarian Imagination*. Albany, NY: SUNY Press.
Kurzman, C. (2011) *The Missing Martyrs: Why There Are So Few Muslim Terrorists*. New York: Oxford University Press.
Labussière, O. (2011) La norme et le mouvant: élements pour une relecture de l'œuvre de Jean Gottmann. *Géographie et cultures* 72:7–23.
Lacoste, Y. (1986) *Géopolitiques des régions françaises*. Paris: Fayard.
———. (2001) Rivalries for territory. In *From Geopolitics to Global Politics: A French Connection*, ed. J. Lévy. London: Frank Cass.
Landes, D. (1998) *The Wealth and Poverty of Nations*. Cambridge, MA: Harvard University Press.
Larner, W. (2011) Governance. In *Wiley-Blackwell Companion to Human Geography*, ed. J. Agnew and J. Duncan. Oxford: Wiley-Blackwell.
Latour, B. (2004) *Politics of Nature: How to Bring the Sciences into Democracy*. Cambridge, MA: Harvard University Press.
Lattimore, O. (1940) *Inner Asian Frontiers of China*. New York: American Geographical Society.
———. (1945) *Solution in Asia*. Boston: Little, Brown.
———. (1949) *The Situation in Asia*. Boston: Little, Brown.
Le Billon, P., and F. El Khatib. (2004) From free oil to "freedom oil": Terrorism, war and US geopolitics in the Persian Gulf. *Geopolitics* 9 (1):109–37.
Legg, S. (2011) Governance. In *Wiley-Blackwell Companion to Human Geography*, ed. J. Agnew and J. Duncan. Oxford: Wiley-Blackwell.
Le Lannou, M. (1975) La leçon de André Siegfried. *Le Monde*, April 6–7.

Lepore, J. (2010) *The Whites of Their Eyes: The Tea Party's Revolution and the Battle over American History*. Princeton, NJ: Princeton University Press.
Levin, P. S., and D. A. Levin. (2002) The real biodiversity crisis. *American Scientist* 90 (1):6–8.
Lévy, J. (2001) A user's guide to world-spaces. In *From Geopolitics to Global Politics: A French Connection*, ed. J. Lévy. London: Frank Cass.
——, ed. (2008) *L'invention du monde. Une géographie de la mondialisation*. Paris: Les Presses de Sciences Po.
Lewis, B. (1995) *The Middle East: A Brief History of the Last 2,000 Years*. New York: Scribner.
Lewis, L. S. (1993) *The Cold War and Academic Governance: The Lattimore Case at Johns Hopkins*. Albany: State University Press of New York.
Lilla, M. (2001) *The Reckless Mind: Intellectuals in Politics*. New York: New York Review of Books.
Liverman, D. (2009) The geopolitics of climate change: Avoiding determinism, fostering sustainable development. *Climate Change* 6:7–11.
Livingstone, D. N. (1992) *The Geographical Tradition: Episodes in the History of a Contested Enterprise*. Oxford: Blackwell.
——. (2007) Science, site and speech: Scientific knowledge and the spaces of rhetoric. *History of the Human Sciences* 20:71–98.
Lovell, N., ed. (1998) *Locality and Belonging*. London: Routledge.
Lowenthal, D. (2005) Personal communication to L. Muscarà, February 19.
Lowi, M. R., and B. R. Shaw, eds. (2000) *Environment and Security: Discourses and Practices*. London: Macmillan.
Luttwak, E. N. (1990) From geopolitics to geo-economics: Logic of conflict, grammar of commerce. *National Interest* 20:17–23.
——. (2009) *The Grand Strategy of the Byzantine Empire*. Cambridge, MA: Harvard University Press.
MacDonald, F. (2007) Anti-astropolitik—outer space and the orbit of geography. *Progress in Human Geography* 31 (5):592–615.
Macfarlane, A. (2011) It's 2050: Do you know where your nuclear waste is? *Bulletin of the Atomic Scientists* 67 (4):30–36.
Machlis, G. E., and T. Hanson. (2008) Warfare ecology. *Bioscience* 58 (8):729–36.
Mackenzie, W. J. M. (1976) *Political Identity*. London: Penguin.
Mackinder, H. J. (1887) The scope and methods of geography. *Proceedings of the Royal Geographical Society*, n.s., 9:141–60.
——. (1904) The geographical pivot of history. *Geographical Journal* 23:421–37.
——. (1919) *Democratic Ideals and Reality: A Study in the Politics of Reconstruction*. London: Constable.
——. (1943) The round world and the winning of the peace. *Foreign Affairs* 21 (4):595–605.
MacKinnon, D. (2011) Reconstructing scale: Towards a new scalar politics. *Progress in Human Geography* 35:21–36.
MacLeod, G., and M. Goodwin. (1999) Reconstructing an urban and regional political economy: On the state, politics, scale, and explanation. *Political Geography* 18:697–730.

Malley, R., and H. Agha. (2001) Camp David: The tragedy of errors. *New York Review of Books*, August 9, 59–65.
Mamadouh, V. (2008) After van Gogh: The geopolitics of the tsunami relief effort in the Netherlands. *Geopolitics* 13 (2):205–31.
———. (2011) Forum on the 2011 "Arab Spring"—Introduction. *Arab World Geographer* 14 (2):111–15.
Mamdani, M. (2000) *When Victims Become Killers*. Princeton, NJ: Princeton University Press.
Manin, B. (1987) On legitimacy and political deliberation. *Political Theory* 15: 338–68.
Mann, M. (1984) The autonomous power of the state. *European Journal* of *Sociology* 25:185–213.
Margalit, A. (2001) Settling scores. *New York Review of Books*, September 20, 20–24.
Marston, S. (2000) The social construction of scale. *Progress in Human Geography* 24:219–42.
Martin-Nielsen, J. (2010) "This war for men's minds": The birth of a human science in Cold War America. *History of the Human Sciences* 23:131–55.
Massey, D. (1994) *Space, Place and Gender*. Minneapolis: University of Minnesota Press.
Maull, O. (1928) *Politischen Grenzen*. Berlin: Zentral-Verlag.
McGann, J. G. (2007) *Think Tanks and Policy-Making in the US: Academics, Advisors and Advocates*. New York: Routledge.
———. (2011) L'influence grandissante des think tanks américains dans le processus d'élaboration des politiques de sécurité contemporaines. *Revue Internationale et Stratégique* 82 (2):119–26.
Menand, L. (2001) College: The end of the Golden Age. *New York Review of Books*, October 18, 44–47.
Mercier, G. (1995) The geography of Friedrich Ratzel and Paul Vidal de la Blache: A comparative analysis. *Annales de Géographie* 583:211–35.
Merrett, C. (2003) Debating destiny: Nihilism or hope in *Guns, Germs, and Steel*? *Antipode* 35 (4):801–6.
Middleton, R. (2011) Trends in piracy: A global problem with Somalia at the core. Selected briefing papers, Conference on Global Challenges, Regional Responses: Forging a Common Approach to Maritime Piracy, organized by the UAE Ministry of Foreign Affairs in association with DP World, Dubai, April 18–19, Dubai School of Government. http://counterpiracy.ae/background_papers.html (accessed September 5, 2011).
Milanovic, B. (2011) *The Haves and the Have-Nots: A Brief and Idiosyncratic History of Global Inequality*. New York: Basic Books.
Miller, B. A. (2000) *Geography and Social Movements: Comparing Antinuclear Activism in the Boston Area*. Minneapolis: University of Minnesota Press.
Miller, D. (1986) Peter Kropotkin (1842–1921): Mutual aid and anarcho-communism. In *Rediscoveries*, ed. J. A. Hall. Oxford: Clarendon Press.
Miller, D., and S. H. Hashmi, eds. (2001) *Boundaries and Justice: Ethical Perspectives*. Princeton, NJ: Princeton University Press.
Milosz, C. (1955) *The Captive Mind*. New York: Vintage.
Mishra, P. (2002) The Afghan tragedy. *New York Review of Books*, January 17, 43–49.

Miyakawa, Y. (2001) The evolution of the political geography of Japan in the perspective of Ratzel and Gottmann. In *On the Centenary of Ratzel's* Politische Geographie: *Europe between Political Geography and Geopolitics*, ed. M. Antonsich et al. Rome: Memorie della Società Geografica Italiana.

Mofson, P. (1999) Global ecopolitics. In *Reordering the World: Geopolitical Perspectives on the Twenty-first Century*, ed. G. J. Demko and W. B. Wood. 2nd ed. Boulder, CO: Westview.

Mohanty, C. T. (1991) Cartographies of struggle. In *Third World Women and the Politics of Feminism*, ed. C. T. Mohanty et al. Bloomington: Indiana University Press.

Moisy, C. (1997) Myths of the global information village. *Foreign Policy* 107:78–87.

Morley, D., and K. Robins. (1995) *Spaces of Identity: Global Media, Electronic Landscapes and Cultural Boundaries*. London: Routledge.

Muller, M. (2008) Reconsidering the concept of discourse for the field of critical geopolitics: Towards discourse as language and practice. *Political Geography* 27:322–38.

Murphy, A. B. (1993) Linguistic regionalism and the social construction of space in Belgium. *International Journal for the Sociology of Language* 104:49–64.

Muscarà, L. (1996) Innovazione tecnologica, spazio e rappresentazione. *Geotema* 2 (6):46–56.

———. (1998) Jean Gottmann's Atlantic "transhumance" and the development of his spatial theory. *Finisterra: Revista Portuguesa de Geografia* 33:159–72.

———. (2001) Gottmann's geographic glossa. *Geojournal* 52 (4):285–93.

———. (2005a) Geografi, etnicità e confini a Versailles. In *Europa: Vecchi confini e nuove frontiere*, ed, E. Dell'Agnese and E. Squarcina. Turin: UTET.

———. (2005b) Territory as a psychosomatic device: Gottmann's kinetic political geography. *Geopolitics* 10:24–49.

———. (2009) Gottmann, J. In *International Encyclopedia of Human Geography*, ed. R. Kitchin and N. Thrift. Amsterdam: Elsevier Science.

Myers, N. (1993) Environmental refugees in a globally warmed world. *Bioscience* 43:752–61.

Myers, N., R. A. Mittermeier, C. G. Mittermeier, G. A. da Fonseca, J. Kent. (2000) Biodiversity hotspots for conservation priorities. *Nature* 403:853–58.

Nagel, T. (1986) *The View from Nowhere*. New York: Oxford University Press.

Naji, A. B. (2006) *Management of Savagery: The Most Critical Stage through which the Umma Will Pass*. Trans. William McCants. Cambridge, MA: John M. Olin Institute for Strategic Studies, Harvard University / United States Military Academy.

National Commission on Terrorist Attacks upon the United States. (2004) *The 9/11 Commission Report: Final Report of the National Commission on Terrorist Attacks upon the United States*. New York: W. W. Norton.

Newburger, H. B, E. L. Birch, and S. Wachter, eds. (2011) *Neighborhood and Life Chances: How Place Matters in Modern America*. Philadelphia: University of Pennsylvania Press.

Newman, D. (2001) Boundaries, borders, and barriers: Changing geographic perspectives on territorial lines. In *Identities, Borders, Orders: Rethinking International Relations Theory*, ed. M. Albert et al. Minneapolis: University of Minnesota Press.

———. (2002) The geopolitics of peacemaking in Israel-Palestine. *Political Geography* 21:629–46.

Newman, D., and A. Paasi. (1998) Fences and neighbours in the postmodern world: Boundary narratives in political geography. *Progress in Human Geography* 22:186–207.

Nicholls, W. (2009) Place, networks, space: Theorizing the geographies of social movements. *Transactions of the Institute of British Geographers* 34:78–93.

Nijman, J. (1992) The limits of superpower: The United States and the Soviet Union since World War II. *Annals* of the *Association* of *American Geographers* 82:681–95.

Nordas, R., and N. P. Gleditsch. (2007) Climate change and conflict. *Political Geography* 26:627–38.

Novotny, P. (2000) *Where We Live, Work, and Play: The Environmental Justice Movement and the Struggle for a New Environmentalism*. Westport, CT: Praeger.

Nowak, M., with R. Highfield. (2011) *SuperCooperators: Altruism, Evolution, and Why We Need Each Other to Succeed*. New York: Free Press.

Nuttall, S., K. Darian-Smith, and L. Gunnar, eds. (1996) *Text, Theory, Space: Postcolonial Representations and Identity*. London: Routledge.

O'Brien, R. (1992) *Global Financial Integration: The End of Geography*. London: Chatham House.

O'Hagan, J. (2000) A "clash of civilizations"? In *Contending Images of World Politics*, ed. G. Fry and J. O'Hagan. London: Macmillan.

Ohmae, K. (1995) *The End of the Nation-State: The Rise of Regional Economies*. New York: McKinsey.

O'Lear, S. (2001) Azerbaijan: Territorial issues and internal challenges in mid-2001. *Post-Soviet Geography and Economics* 42:305–12.

O'Loughlin, J. (1986) Spatial models of international conflict: Extending current theories of war behavior. *Annals of the Association* of *American Geographers* 76:63–80.

———. (2003) Spatial analysis in political geography. In *A Companion to Political Geography*, ed. J. Agnew, K. Mitchell, and G. Toal. Oxford: Blackwell.

O'Loughlin, J., and G. Ó Tuathail. (2009) Accounting for separatist sentiment in Bosnia-Herzegovina and the North Caucasus: A comparative analysis of survey responses. *Ethnic and Racial Studies* 32:591–615.

O'Loughlin, J., and C. Raleigh. (2008) Spatial analysis of civil war violence. In *The Sage Handbook of Political Geography*, ed. K. R. Cox et al. London: Sage.

O'Loughlin, J., and H. Van der Wusten. (1990) Political geography of pan-regions. *Geographical Review* 80:1–20.

O'Loughlin, J., and F. D. W. Witmer. (2011) The localized geographies of violence in the North Caucasus of Russia, 1999–2007. *Annals of the Association of American Geographers* 101:178–201.

Osei-Kwame, P., and P. J. Taylor. (1984) A politics of failure: The political geography of Ghanaian elections, 1954–1979. *Annals of the Association of American Geographers* 74:574–89.

O'Sullivan, P. (1986) *Geopolitics*. New York: St. Martin's.

Ó Tuathail, G. (1993) The effacement of place? US foreign policy and the Gulf crisis. *Antipode* 25:4–31.

———. (1996) *Critical Geopolitics*. Minneapolis: University of Minnesota Press.

——. (2000a) The postmodern geopolitical condition: States, statecraft, and security at the millennium. *Annals of the Association of American Geographers* 90:166–78.
——. (2000b) Spiritual geopolitics: Fr Edmund Walsh and Jesuit anti-communism. In *Geopolitical Traditions: A Century of Geopolitical Thought*, ed. K. Dodds and D. Atkinson. London: Routledge.
——. (2003) "Just out looking for a fight": American affect and the invasion of Iraq. *Antipode* 35:857–70.
——. (2009) Placing blame: Making sense of Beslan. *Political Geography* 28:4–15.
——. (2010) Localizing geopolitics: Disaggregating violence and return in conflict regions. *Political Geography* 29:256–65.
Ó Tuathail, G., and J. A. Agnew. (1992) Geopolitics and discourse: Practical geopolitical reasoning in American foreign policy. *Political Geography Quarterly* 11:190–204.
Ó Tuathail, G., and J. O'Loughlin. (2009) After ethnic cleansing: Return outcomes in Bosnia-Herzegovina a decade beyond war. *Annals of the Association of American Geographers* 99:1045–53.
Paasi, A. (1995) *Territories, Boundaries, and Consciousness: The Changing Geographies of the Finnish-Russian Border*. Chichester, UK: John Wiley.
——. (2004) Place and region: Looking through the prism of scale. *Progress in Human Geography* 28 (24):536–46.
Packer, G. (2006) Knowing the enemy: Can social scientists redefine the "war on terror"? *New Yorker*, December 18, 60–69.
Paddison, R. (1983) *The Fragmented State: The Political Geography of Power*. Oxford: Blackwell.
Painter, J. (2008) Geographies of space and power. In *The Sage Handbook of Political Geography*, ed. K. R. Cox et al. London: Sage.
Panofsky, E. (1969) *Renaissance and Renascences in Western Art*. New York: Harper and Row.
Parker, G. (1998) *Geopolitics: Past, Present, and Future*. Washington, DC: Pinter.
——. (2001) Ratzel, the French school and the birth of alternative geopolitics. In *On the Centenary of Ratzel's* Politische Geographie: *Europe between Political Geography and Geopolitics*, ed. M. Antonsich et al. Rome: Memorie della Società Geografica Italiana. Also with the same title in *Political Geography* 19:957–69.
Parker, W. H. (1982) *Mackinder: Geography as an Aid to Statecraft*. Oxford: Oxford University Press.
Pearcy, G. E., R. H. Fifield, et al. (1948) *World Political Geography*. New York: Crowell.
Peet, R., and M. Watts, eds. (1996) *Liberation Ecologies: Environment, Development, Social Movements*. London: Routledge.
Penck, A. (1916) Der Krieg und das Studium der Geographie. *Zeitschrift der Gesellschaft Fur Erdkunde zu Berlin*: 159–76 and 222–48.
Perlmutter, A. (1997) *Making the World Safe for Democracy: A Century of Wilsonianism and Its Totalitarian Challengers*. Chapel Hill: University of North Carolina Press.
Perrin, N. (1979) *Giving Up the Gun: Japan's Reversion to the Sword, 1543–1879*. Boulder, CO: Shambhala.
Pettis, M. (2001) *The Volatility Machine: Emerging Economies and the Threat of Financial Collapse*. New York: Oxford University Press.

Piore, M., and C. Sabel. (1984) *The Second Industrial Divide*. New York: Basic Books.
Pomeranz, K. (2009) The Great Himalayan Watershed: Agrarian crisis, mega-dams and the environment. *New Left Review* 58:5–39.
Powell, R. C. (2008) Configuring an Arctic commons? *Political Geography* 27:827–32.
———. (2010) Lines of possession? The anxious constitution of a polar geopolitics. *Political Geography* 29:74–77.
Pred, A. (1990) *Making Histories and Transforming Human Geographies: The Local Transformation of Practice, Power Relations and Consciousness*. Boulder, CO: Westview.
Preda, A. (2009) *Framing Finance: The Boundaries of Markets and Modern Capitalism*. Chicago: University of Chicago Press.
Prévélakis, G. (1996) La notion du territoire dans la pensée de Jean Gottmann. *Géographie et Cultures* 20:81–92.
Price, M. D. (1999) Nongovernmental organizations on the geopolitical front line. In *Reordering the World: Geopolitical Perspectives on the Twenty-first Century*, ed. G. J. Demko and W. B. Wood. 2nd ed. Boulder, CO: Westview.
Princen, T., and M. Finger. (1994) *Environmental NGOs in World Politics: Linking the Local and the Global*. New York: Routledge.
Purcell, M. (2008) *Recapturing Democracy: Neoliberalization and the Struggle for Alternative Urban Futures*. London: Routledge.
Quinn, D. (1997) The correlates of change in international financial regulation. *American Political Science Review* 91:531–51.
Raffestin, C. (1988) Postface in *Géographie politique*, by F. Ratzel. Geneva: Editions Régionales Européennes.
———. (2001) From text to image. In *From Geopolitics to Global Politics: A French Connection*, ed. J. Lévy. London: Frank Cass.
Rashid, A. (2000) *Taliban: Militant Islam, Oil and Fundamentalism in Central Asia*. New Haven, CT: Yale University Press.
Ratzel, F. (1896) Die Gesetze des räumlichen Wachstums der Staaten. *Petermanns Mitteilungen* 42:97–107.
———. (1897) *Politische Geographie*. Munich: R. Oldenbourg.
———. (1923) *Politische Geographie*. 3rd ed. Munich: R. Oldenbourg.
———. (1969) The laws of the spatial growth of states. In *The Structure of Political Geography*, ed. R. E. Kasperson and J. V. Minghi. Chicago: Aldine.
Rawls, J. (1971) *A Theory of Justice*. Cambridge, MA: Harvard University Press.
Reclus, E. (1905–1908) *L'homme et la terre*. Paris: Librairie Universelle.
Reifer, T. E. (2006) Militarization, globalization, and Islamist social movements: How today's ideology of Islamophobia fuels militant Islam. *Human Architecture: Journal of the Sociology of Self-Knowledge* 5 (1):51–72.
Retaille, D. (2001) Geopolitics in history. In *From Geopolitics to Global Politics: A French Connection*, ed. J. Lévy. London: Frank Cass.
Robic, M.-C. (1994) National identity in Vidal's *Tableau de la géographie de la France*: From political geography to human geography. In *Geography and National Identity*, ed. D. Hooson. Oxford: Blackwell.
Robin, R. (2001) *The Making of the Cold War Enemy: Culture and Politics in the Military-Intellectual Complex*. Princeton, NJ: Princeton University Press.

Robinson, F. (1999) *Globalizing Care: Ethics, Feminist Theory, and International Relations*. Boulder, CO: Westview.
Rodgers, D. T. (2011) *Age of Fracture*. Cambridge, MA: Harvard University Press.
Rogers, P. (2000) Resource issues. In *Issues in International Relations*, ed. T. C. Salmon. Boulder, CO: Westview.
Rokkan, S. (1980) Territories, centres, and peripheries: Toward a geoethnic-geoeconomic-geopolitical model of differentiation within Western Europe. In *Centre and Periphery: Spatial Variation in Politics*, ed. J. Gottmann. London: Sage.
Rosenberg, E. S. (1982) *Spreading the American Dream: American Economic and Cultural Expansion, 1890–1945*. New York: Hill & Wang.
Routledge, P. (1992) Putting politics in its place: Baliapal, India, as a terrain of resistance. *Political Geography* 11:588–611.
———. (1994) *Terrains of Resistance: Non-violent Social Movements and the Contestation of Place in India*. Westport, CT: Praeger.
Rubin, C. T. (1998) *The Green Crusade: Rethinking the Roots of Environmentalism*. Lanham, MD: Rowman & Littlefield.
Sachs, J. D. (2000) Tropical underdevelopment. Paper presented at the annual meeting of the Economic History Association, Los Angeles, September 8.
———. (2005) *The End of Poverty*. New York: Penguin.
Sack, R. D. (1986) *Human Territoriality: Its Theory and History*. Cambridge: Cambridge University Press.
———. (2003) *A Geographical Guide to the Real and the Good*. London: Routledge.
Sage, D. (2008) Framing space: A popular geopolitics of American Manifest Destiny in outer space. *Geopolitics* 13:27–53.
Sahlins, P. (1989) *Boundaries: The Making of France and Spain in the Pyrenees*. Berkeley: University of California Press.
Said, E. (1978) *Orientalism*. New York: Vintage.
———. (2000) Palestinians under siege. *London Review* of *Books*, December 14, 9–14.
Saikal, A. (2000) "Islam and the West"? In *Contending Images of World Politics*, ed. G. Fry and J. O'Hagan. London: Macmillan.
Sandner, G. (1994) In search of identity: German nationalism and geography, 1871–1910. In *Geography and National Identity*, ed. D. Hooson. Oxford: Blackwell.
Sanguin, A.-L. (1985) André Siegfried: An unconventional French political geographer. *Political Geography Quarterly* 4:79–83.
———. (1996) Jean Gottmann (1915–1994) et la géographie politique. In *La Géographie française à l'époque classique (1918–1968)*, ed. P. Claval and A.-L. Sanguin. Paris: L'Harmattan.
———. (2010) *André Siegfried, un visionnaire humaniste entre géographie et politique*. Paris: L'Harmattan.
Sanguin, A.-L., and G. Prévélakis. (1996) Jean Gottmann (1915–1994) un pionnier de la géographie politique. *Annales de Géographie* 105:73–78.
Sassen, S. (1991) *The Global City*. Princeton, NJ: Princeton University Press.
Satterthwaite D. (2008) Cities' contribution to global warming: Notes on the allocation of greenhouse gas emissions. *Environment and Urbanization* 20:539–49.
Schell, J. (2011) Hiroshima to Fukushima. *Nation* 292 (14):6–7.
Schmitt, C. (1996) *The Concept of the Political*. Chicago: University of Chicago Press.

Schwarz, M., and M. Thompson. (1990) *Divided We Stand: Redefining Politics, Technology and Social Choice*. Philadephia: University of Pennsylvania Press.
Scott, A. J. (1998) *Regions and the World Economy: The Coming Shape of Global Production, Competition, and Political Order*. Oxford: Oxford University Press.
Scott, J. W. (1992) Experience. In *Feminists Theorize the Political*, ed. J. Butler and J. W. Scott. New York: Routledge.
Sen, A. (2006) *Identity and Violence: The Illusion of Destiny*. New York: W. W. Norton.
Shapiro, M. J. (1999) Samuel Huntington's moral geography. *Theory and Event* 2 (4):1–11.
Sharp, J. P. (2000) *Condensing the Cold War:* Reader's Digest *and American Identity*. Minneapolis: University of Minnesota Press.
———. (2002) Gender in a political and patriarchal world. In *Handbook of Cultural Geography*, ed. K. Anderson et al. London: Sage.
———. (2003) Feminist and postcolonial engagements. In *A Companion to Political Geography*, ed. J. Agnew, K. Mitchell, and G. Toal. Oxford: Blackwell.
Shatz, A. (2011) Is Palestine next? *London Review of Books*, July 14, 8–14.
Shaxson, N. (2011) *Treasure Islands: Tax Havens and the Men Who Stole the World*. London: Bodley Head.
Shin, M. E., and Agnew, J. A. (2008) *Berlusconi's Italy: Mapping Contemporary Italian Politics*. Philadelphia: Temple University Press.
Sidaway, J. D. (2009) Overwriting geography: Mackinder's presences, a dialogue with David Hooson. *Geopolitics* 14 (1):163–70.
Siegfried, A. (1913) *Tableau de la France de l'Ouest sous la Troisième République*. Paris: Armand Colin.
———. (1930) *France, a Study in Nationality*. New Haven, CT: Yale University Press. (In French as [1930] *Tableau des partis en France*. Paris: Grasset.)
Silvern, S. E. (1999) Scales of justice: Law, Indian treaty rights and the political construction of scale. *Political Geography* 18:639–68.
Sinclair, T. J. (2005) *The New Masters of Capital: American Bond Rating Agencies and the Politics of Creditworthiness*. Ithaca, NY: Cornell University Press.
Skocpol, T. (1994) *Social Revolutions in the Modern World*. Cambridge: Cambridge University Press.
Slater, D. (1999) Situating geopolitical representations: Inside/outside and the power of imperial interventions. In *Human Geography Today*, ed. D. Massey et al. Cambridge: Polity Press.
Sluyter, A. (2003) Neo-environmental determinism, intellectual damage control and nature/society science. *Antipode* 35:813–17.
Smith, D. M. (1999) Geography and ethics. How far should we go? *Progress in Human Geography* 23:119–25.
Smith, N. (1994) Shaking loose the colonies: Isaiah Bowman and the "decolonization" of the British Empire. In *Geography and Empire*, ed. A. Godlewska and N. Smith. Oxford: Blackwell.
———. (2003) *American Empire: Roosevelt's Geographer and the Prelude to Globalization*. Berkeley: University of California Press.
Smith, W. D. (1986) *The Ideological Origins of Nazi Imperialism*. New York: Oxford University Press.

Solomon, S. (2010) *Water: The Epic Struggle for Wealth, Power, and Civilization*. New York: HarperCollins.
Sontag, D. (2001) Quest for Mideast peace: How and why it failed. *New York Times*, July 26, A1–12.
Soros, G. (1998–1999) Capitalism's last chance? *Foreign Policy* 113:55–66.
Sprout, H., and M. Sprout. (1939) *The Rise of American Naval Power*. Princeton, NJ: Princeton University Press.
——. (1943) *Toward a New Order of Sea Power*. Princeton, NJ: Princeton University Press.
——. (1962) *Foundations of International Politics*. New York: Van Nostrand.
——. (1965) *The Ecological Perspective on Human Affairs, with Special Reference to International Politics*. Princeton, NJ: Princeton University Press.
——. (1978) *The Context of Environmental Politics: Unfinished Business for America's Third Century*. Lexington: University Press of Kentucky.
Spykman, N. J. (1944) *The Geography of the Peace*. New York: Harcourt Brace.
Staeheli, L. (1994) Empowering political struggle: Spaces and scales of resistance. *Political Geography* 13:387–91.
Stanley, M. (1978) *The Technological Conscience: Survival and Dignity in an Age of Expertise*. Chicago: University of Chicago Press.
Starn, R. (1994) *Ambrogio Lorenzetti: The Palazzo Pubblico, Siena*. New York: George Braziller.
Stephenson, S. R., L. C. Smith, and J. A. Agnew. (2011) Divergent long-term trajectories of human access to the Arctic. *Nature Climate Change* 1:156–60.
Stern, A. (1975) Political legitimacy in local politics: The Communist Party in northeastern Italy. In *Communism in Italy and France*, ed. D. Blackmer and S. Tarrow. Princeton, NJ: Princeton University Press.
Stern, R. J. (2010) United States cost of military force projection in the Persian Gulf, 1976–2007. *Energy Policy* 38:2816–25.
Stewart, J. Y. (2001) Our town: Life in the killing zone. *Los Angeles Times Magazine*, January 7, 10–17, 35.
Stiglitz, J. E. (2008) The fruit of hypocrisy. *Guardian*, September 16, 30.
——. (2010) *Freefall: America, Free Markets, and the Sinking of the World Economy*. New York: W. W. Norton.
Swyngedouw, E. (2010) Apocalypse forever: Post-political populism and the spectre of climate change. *Theory, Culture & Society* 27:213–32.
Takeuchi, K. (1994) The Japanese imperial tradition, western imperialism and modern Japanese geography. In *Geography and Empire*, ed. A. Godlewska and N. Smith. Oxford: Blackwell.
Tarrow, S. (1994) *Power in Movement: Social Movements, Collective Action and Politics*. Cambridge: Cambridge University Press.
Taylor, P. J. (1989) *Political Geography: World-Economy, Nation-State and Locality*. London: Longman.
——. (1994) The state as container: Territoriality in the modern world-system. *Progress in Human Geography* 18:151–62.
——. (2003) Radical political geographies. In *A Companion to Political Geography*, ed. J. Agnew, K. Mitchell, and G. Toal. Oxford: Blackwell.

Tesini, M. (1986) *Oltre la città rossa. L'alternativa mancata di Dossetti a Bologna (1956–58)*. Bologna: II Mulino.
Thaa, W. (2001) "Lean citizenship": The fading away of the political in transnational democracy. *European Journal of International Relations* 7:503–23.
Thomas, D. C. (2005) Human rights ideas, the demise of Communism, and the end of the Cold War. *Journal of Cold War Studies* 7:110–41.
Thomson, J. E. (1994) *Mercenaries, Pirates, and Sovereigns: State-Building and Extra-Territorial Violence in Early Modern Europe*. Princeton, NJ: Princeton University Press.
Thrift, N. (2000) It's the little things. In *Geopolitical Traditions: A Century of Geopolitical Thought*, ed. K. Dodds and D. Atkinson. London: Routledge.
Thrift, N., and S. Pile, eds. (1995) *Mapping the Subject: Geographies of Cultural Transformation*. London: Routledge.
Thual, F. (1995) *Les conflits identitaires*. Paris: Ellipses.
Tilly, C. (1986) *The Contentious French*. Cambridge, MA: Harvard University Press.
———. (1990) *Coercion, Capital and European States: AD 990–1992*. Oxford: Blackwell.
Tilly, C., and W. P. Blockmans, eds. (1994) *Cities and the Rise of States in Europe, A.D. 1000 to 1800*. Boulder, CO: Westview.
Toal, G., and C. T. Dahlman. (2011) *Bosnia Remade: Ethnic Cleansing and Its Reversal*. Oxford: Oxford University Press.
Tollefson, J. W. (1993) *The Strength Not to Fight*. Boston: Little, Brown.
UNODC. (2010) *World Drug Report 2010*. New York: United Nations Office on Drug Control.
Van Valkenburg, S. (1939) *Elements of Political Geography*. New York: Prentice-Hall.
Verger, F. ed. (2002) *L'espace, nouveau territoire. Atlas des satellites et des politiques spatiales*. Paris: Belin.
Verhoeven, H. (2009) The self-fulfilling prophecy of failed states: Somalia, state collapse and the Global War on Terror. *Journal of East African Studies* 3:405–25.
Vidal de la Blache, P. (1903) *Tableau de la géographie de la France*. Paris: Hachette.
———. (1917) *La France de l'Est: Lorraine-Alsace*. Paris: Armand Colin.
Viroli, M. (1995) *For Love of Country: An Essay on Patriotism and Nationalism*. Oxford: Oxford University Press.
Von Neumann, J. (1955) Can we survive technology? *Fortune*, June, 106–8, 152–53.
Wacquant, L. J. D. (1994) The new urban color line: The state and the fate of the ghetto in post-Fordist America. In *Social Theory and the Politics of Identity*, ed. C. Calhoun. Oxford: Blackwell.
Wade, R. (1998–1999) The coming fight over capital flows. *Foreign Policy* 113:41–54.
Waldoff, S. T., and A. A. Fawcett. (2011) Can developed economies combat dangerous anthropogenic climate change without near-term reductions from developing countries? *Climatic Change* 107:635–41.
Wallerstein, I. (1974) *The Modern World-System: Capitalist Agriculture and the Origins of the European World-Economy in the Sixteenth Century*. New York: Academic Press.
———. (1993) *Geopolitics and Geoculture*. Cambridge: Cambridge University Press.
Walzer, M. (1997) *On Toleration*. New Haven, CT: Yale University Press.
Watson Institute. (2011) *The Costs of War since 2001: Iraq, Afghanistan, and Pakistan*. Providence, RI: Watson Institute, Brown University.
Watts, D. J. (2011) *Everything Is Obvious: *Once You Know the Answer*. New York: Crown.

Watts, M. J. (1999) Collective wish images: Geographical imaginaries and the crisis of national development. In *Human Geography Today*, ed. D. Massey et al. Cambridge: Polity Press.

———. (2004) Resource curse? Governmentality, oil and power in the Niger Delta, Nigeria. *Geopolitics* 9 (1):50–80.

Weber, E. (1976) *Peasants into Frenchmen: The Modernization of Rural France, 1870–1914*. Stanford, CA: Stanford University Press.

Weigert, H. J, H. Brodie, E. W. Doherty, J. R. Fernstrom, E. Fischer, and D. Kirk. (1957) *Principles of Political Geography*. New York: Appleton Century Crofts.

Whittlesey, D. S. (1939) *The Earth and the State*. New York: Henry Holt.

Williams, A. J. (2007) Hakumat al Tayarrat: The role of air power in the enforcement of Iraq's borders. *Geopolitics* 12:505–28.

———. (2010) A crisis in aerial sovereignty? Considering the implications of recent military violations of national airspace. *Area* 42:51–59.

Williams, C. H. (1989) The question of national congruence. In *A World in Crisis? Geographical Perspectives*, ed. R. J. Johnston and P. J. Taylor. Oxford: Blackwell.

Williams, R. W. (1999) Environmental injustice in America and its politics of scale. *Political Geography* 18:49–73.

Wilson, E. O. (2002) Hotspots: Preserving pieces of a fragile biosphere. *National Geographic*, January, 86–89.

Wittfogel, K. (1929) Geopolitik, geographischer Materialismus und Marxismus. *Unter den Banner des Marxismus* 3 (1, 4, 5). Translated as Wittfogel (1985) below.

———. (1957) *Oriental Despotism: A Comparative Study of Total Power*. New Haven, CT: Yale University Press.

———. (1985) Geopolitics, geographical materialism, and Marxism. *Antipode* 17:21–72.

Wokler, R. (1987) Saint Simon and the passage from political to social science. In *The Languages of Political Theory in Early-Modern Europe*, ed. A. Pagden. Cambridge: Cambridge University Press.

Wolf, A. T., and J. H. Hamner. (2000) Trends in transboundary water disputes and dispute resolution. In *Environment and Security: Discourses and Practices*, ed. M. R. Lowi and B. R. Shaw. London: Macmillan.

Wolf, E. R. (1982) *Europe and the People without History*. Berkeley: University of California Press.

The World Bank. (2011) "The World Development Report 2011: Conflict, Security and Development." Washington, DC: The International Bank for Reconstruction and Development/The World Bank.

Yack, B. (2001) Popular sovereignty and nationalism. *Political Theory* 29:517–36.

Yiftachel, O., and H. Yacobi. (2002) Urban ethnocracy: Ethnicization and production of space in an Israeli "mixed city." *Environment and Planning D: Society and Space* 21:673–93.

Young, I. M. (1987) Impartiality and the civic public: Some implications of feminist critiques of moral and political theory. In *Feminism as Critique: On the Politics of Gender*, ed. S. Benhabib and D. Cornell. Oxford: Blackwell.

———. (1990) *Justice and the Politics of Difference*. Princeton, NJ: Princeton University Press.

Young, O. (2009) The Arctic in play: Governance in a time of rapid change. *International Journal of Marine and Coastal Law* 24:423–42.
Zakaria, F. (2008) *The Post-American World*. New York: W. W. Norton.
———. (2011) *The Post-American World, Release 2.0*. New York: W. W. Norton.
Ziang, C. (2011) Why do people misunderstand climate change? Heuristics, mental models and ontological assumptions. *Climatic Change* 108 (1–2):31–46.
Zuckerman, S. (1982) *Nuclear Illusion and Reality*. New York: Viking.

Index

Note: Figures and tables are indicated by "f" and "t," respectively, following page numbers.

About the Authors

John Agnew is Distinguished Professor of Geography at UCLA. His publications include *Hegemony: The New Shape of Global Power* (2005) and *Globalization and Sovereignty* (2009).

Luca Muscarà is associate professor of geography at the University of Molise. He is the former editor of *Sistema Terra* and has written widely on the history of political geography.